STUDY GUIDE

TO ACCOMPANY

GEOGRAPHY:
REALMS, REGIONS, AND CONCEPTS

FIFTEENTH EDITION

H. J. DE BLIJ
Michigan State University

PETER O. MULLER
University of Miami

JAN NIJMAN
University of Miami

Prepared by

PETER O. MULLER
ELIZABETH MULLER HAMES

WILEY

JOHN WILEY & SONS

Cover photographs by Zachary Woodward; © Jan Nijman

ISBN   978-1-118-16685-7

Printed in the United States of America

10 9 8 7 6 5 4 3 2 1

Printed and bound by Bind-Rite/Robbinsville

A FOREWORD TO STUDENTS

This **Study Guide** is designed to enhance your learning of world regional geography, building upon your parallel experiences in the classroom and in reading the chapters of the textbook. It is recommended that you have an *atlas* handy; the base maps in the textbook will readily suffice if no other atlas is available. The only other resources you will need to complete the exercises in this manual are a set of colored pencils or pens and (optional) tracing paper.

One of the aims of this **Study Guide** is to get you to become more familiar with maps. For each world geographic realm, we have provided four outline maps (located at the end of each chapter) on which you are asked to locate and enter important geographic information. The essay questions in each chapter are also attuned to maps—many, in fact, cry out for a cartographic answer. That is exactly what you should consider doing: use sketch maps whenever the map is the best way to explain what you mean (like a picture, a map is often worth at least a thousand words). You can use tracings of the printed maps in order to get the necessary outlines of the geographic realms; it is also a good idea to use such tracings in answering map questions as a sort of "test run" before placing your final answers on the printed map. Moreover, placing information on maps as you read the textbook can be a valuable technique for later reviewing before examinations; an extra printed outline map is always provided in each chapter of this manual.

Each chapter of this **Study Guide** follows an identical format. The opening section describes the *objectives* of the chapter together with a list of things you should be able to do once you have learned the content of the parallel textbook chapter. This is followed by a *glossary* of terms covered in the book chapter, including appropriate terminology from the main textbook glossary as well as additional definitions; the **Study Guide** glossary is arranged in the order in which terms appear in the book, and carries page references to the text. Then comes the main study and review section, a series of detailed, side-by-side *questions and answers* for you to sharpen your understanding of the high points of the subject matter. *Map exercises* appear next, divided into two sections: (1) a series of map-comparison queries arranged around the textbook maps, and (2) three assignments to fill in basic physical, political-cultural, and economic-urban geographic information for each realm; as indicated above, a fourth blank map is included for your own use. Finally, comes a *practice examination* consisting of short answers (multiple choice, true-false, fill-ins, and matching) and some longer essay questions—an answer key for the short-answer questions is found at the back of this manual.

You should find these exercises helpful in this course, particularly for personal studying purposes. They will also assist you to learn basic world geography and to sharpen your map interpretation skills. Such knowledge is not only important as a contribution to your becoming a liberally-educated person—it may already be vital for success in the ever more globalized economy and society of which we are now all a part.

A FOREWORD TO INSTRUCTORS

The aims of this Student Supplement to the fifteenth edition of *Geography: Realms, Regions, and Concepts* are spelled out in the Student Foreword above.

This student **Study Guide** is designed to complement the text in several ways, with each chapter providing a lengthy glossary, a detailed list of study questions-and-answers, map exercises that require comparisons and placement of spatial information on outline maps, and practice examination questions. You should also be aware of the availability of the other ancillaries listed in the textbook's Preface, which contain a number of items pertinent to the teaching of a world regional geography course based on this textbook/integrated learning system.

ACKNOWLEDGMENTS

We are most grateful to a number of people for their assistance and advice during the preparation of this **Study Guide**. The maps were designed by Mapping Specialists, Inc. of Fitchburg, Wisconsin, and supervised by Don Larson. At Wiley we are most thankful for the support and assistance of Darnell Sessoms, Veronica Armour, and Ryan Flahive. We also acknowledge again those who helped in the preparation of previous editions of this **Study Guide** to accompany *The World Today: Concepts and Regions in Geography*.

Peter O. Muller
Elizabeth Muller Hames
Coral Gables, Florida
August 2, 2011

TABLE OF CONTENTS

INTRODUCTION

WORLD REGIONAL GEOGRAPHY

OBJECTIVES OF THIS CHAPTER

The Introduction is an overview of world regional geography. Proceeding from a discussion of basic concepts of regionalism, the topics of scale, the natural environment, culture, landscape, and global population patterns are explored. Additional human-geographic variables are introduced, notably political and economic forces (the latter are explored in some depth within the context of economic development), and provide the basis for establishing the framework of world geographic realms to be pursued in this book. The chapter concludes with a discussion of the relationship between regional and systematic (topical) geography.

Having learned this basic background to the subject, you should be able to:

1. Differentiate among the various basic concepts of realms and regions, and understand the nature of regional structure.

2. Understand the general aspects of the natural environment and its evolution over the past few million years.

3. Explain broad, world-scale patterns of terrain, hydrography (water distribution), and climate, with some emphasis on the interrelationship of these environmental systems.

4. Understand the notion of culture and how it is expressed to form human landscapes in time and space.

5. Identify and discuss the global distribution of humankind, with emphasis on the leading population concentrations.

6. Understand the importance of influential developed areas and how, at the global scale, the economically advantaged countries function as a core that dominates the less developed countries that constitute a "have-not," or economically-disadvantaged periphery.

7. Name, map, and differentiate among the 12 world geographic realms that form the spatial framework to be followed in the remaining chapters of the book.

GLOSSARY

Cartography (5)

Map making.

Spatial perspective (7)

Pertaining to space on the Earth's surface; synonym for *geographic(al)*.

Scale (7)

Representation of a real-world phenomenon at a certain level of reduction or generalization. In cartography, the ratio of map distance to actual ground distance.

Geographic realm (8-9)

The basic spatial unit in our world regionalization scheme. Each realm is defined in terms of a synthesis of its total human geography–a composite of its leading cultural, economic, historical, political, and appropriate environmental features.

Transition zone (9)

An area of spatial change where the peripheries of two adjacent realms or regions join; marked by a gradual shift (rather than a sharp break) in the characteristics that distinguish these neighboring geographic entities from one another.

Region (11-12)

An area of the Earth's surface marked by specific criteria. A commonly used term and a geographic concept of central importance.

Regional boundaries (12)

In theory, the line that circumscribes a region. But razor-sharp lines are seldom encountered, even in nature (e.g., a coastline constantly changes depending upon the tide).

Absolute and Relative Location (12)

The position of a place or region on the Earth's surface. *Absolute location* is that position expressed in degrees, minutes, and seconds of latitude and longitude; *relative location* is that position relative to the position of other places and regions. Thus Palmyra, New Jersey is located at 40^0 North latitude, 75^0 West longitude; relatively speaking, this suburban community is located on the eastern bank of the Delaware River facing Tacony, a factory neighborhood in the Lower Northeast portion of the city of Philadelphia, Pennsylvania.

Formal region (12)

A type of region marked by a certain degree of homogeneity in one or more phenomena, such as physical or cultural properties, or both.

Functional region (12)

A region marked less by its sameness than its dynamic internal structure, formed by a set of places and their interactions.

Spatial system (12)

Any group of objects or institutions and their mutual interactions; geography treats systems that are expressed spatially–such as regions. *Functional regions* are spatial systems.

Hinterland (12)

The surrounding area served by an urban center. This area is both served by the urban central core and comes under its cultural and economic influence.

Natural landscape (13)

The array of landforms that constitutes the Earth's surface (mountains, hills, plains, and plateaus) and the physical features that marks them (such as water bodies, soils, and vegetation). Each geographic realm has its distinctive combination of natural landscapes.

Continental drift (12)

The slow movement of continents controlled by the processes associated with *plate tectonics*.

Plate tectonics (13-14)

Bonded portions of the Earth's mantle and crust, called *tectonic plates*, averaging 60 miles (100 kilometers) in thickness. More than a dozen such plates exist (see Fig. G-5), most of continental proportions, and they are in motion. Where they meet one slides under the other, crumpling the surface crust and producing significant volcanic and seismic (earthquake) activity. A major mountain building force.

Pacific Ring of Fire (14)

Zone of crustal instability along tectonic plate boundaries, marked by earthquakes and volcanic activity, which rings the Pacific Ocean basin.

Ice age (15)

A stretch of geologic time during which the Earth's average atmospheric temperature is lowered; causes the equatorward expansion of continental icesheets in the high latitudes and the growth of mountain glaciers in and around the highlands of the lower latitudes.

Pleistocene epoch (15)

Current period of geologic time that spans the rise of humanity, beginning about 2 million years ago. Marked by *glaciations* (repeated advances of continental-scale ice-sheets) and milder interglaciations (ice withdrawals). Although the last 10,000 years are known as the Holocene, Pleistocene-like conditions seem to be continuing and the present is likely to be another Pleistocene interglacial; the glaciers will return.

Glaciations (15)

Colder phases within an ice age.

Interglacials (15)

Warmer phases within an ice age in which ice recedes poleward.

Weather (16)

The immediate state of the atmosphere.

Climate (16)

The long-term conditions (over at least 30 years) of aggregate weather over a region, summarized by averages and measures of variability; a synthesis of the succession of weather events we have learned to expect at any given location.

Population distribution (19)

The way people have arranged themselves in geographic space. One of human geography's most essential expressions because it represents the sum total of the adjustments that a population has made to its natural, cultural, and economic environments.

Population density (19)

The number of people per unit area.

Urbanization (20)

A term with several connotations. The proportion of a country's population living in cities and towns (metropolitan areas) is its level of urbanization.

Cultural landscape (21)

The composite of human imprints on the surface of the Earth. By this progressive imprinting of the human presence, the physical landscape is modified into the cultural landscape, forming an interacting unity between the two. May also be regarded as a composition of artificial spaces that serves as background for the collective human experience.

State (23)

A politically organized territory that is administered by a sovereign government and is recognized by a significant portion of the international community. A state must also contain a permanent resident population, an organized economy, and a functioning internal circulation system.

European state model (21)

A state consisting of legally defined territory, inhabited by a population governed from a capital city by a representative government.

Sovereignty (23)

Legal concept meaning that the government of a state rules supreme within its borders.

Economic geography (28)

The field of geography that focuses on the diverse ways in which people earn a living, and how the goods and services they produce are expressed and organized spatially.

Development (30)

The economic, social, and institutional growth of national states. This can be measured in a number of ways; the text uses the fourfold scheme for ranking countries according to national income devised by the World Bank (Fig. G-11; pp. 28-29). Other leading development indicators are: national product per person, labor-force occupational structure, productivity per worker, per capita energy consumption, transportation/communications facilities per person, manufactured-metals consumption per person, and such rates as literacy, per capita savings, and individual caloric intake.

Core area (31)

In geography, a term with several connotations. *Core* refers to the center, heart, or focus. The core area of a state is constituted by the national heartland, the largest population cluster, the most productive region, and the part of the country with the greatest centrality and accessibility–probably containing the capital city as well.

Regional disparity (31)

The spatial unevenness in standard of living that occurs within a country, whose "average," overall income statistics invariably mask the differences that exist between the extremes of the wealthy core and the poor periphery.

Globalization (32)

The gradual reduction of regional contrasts resulting from increasing international cultural, economic, and political exchanges.

Gross National Income (GNI) (33)

The total income earned from all gods and services produced by the citizens of a country, within or outside of its borders, during a calendar year.

Systematic geography (37)

Topical geography: cultural, political, economic geography, and the like.

SELF-TESTING QUESTIONS

Cover the right side of the page with a sheet of paper. Uncover each line after you have attempted to answer the question in the left column. If necessary, refer to textbook page(s) listed at the right.

Question	Answer	Page
Regional Concepts		
Define *spatial*.	Pertaining to space on the Earth's surface; synonym for "geographic(al)."	7
Concepts of Scale		
What is *scale*?	Broadly, the level of generalization; specifically, the ratio of map distance to actual ground distance.	7
What characterizes each major world realm?	A special combination of cultural, physical, historical, economic, and organizational properties.	7-8
How do most regional boundaries appear in the landscape?	As transition zones rather than razor-sharp lines.	9
Explain how regionalization is the geographer's means of classification.	According to selected criteria, specific meanings are assigned to certain areas.	11-12
Why is *relative location* a more practical measure than *absolute location*?	*Absolute location* refers only to latitude and longitude, whereas *relative location* refers to a place in terms of its position relative to other regions.	12
How does a *formal region* differ from a *functional region*?	*Formal regions* are marked by internal homogeneity; *functional regions* are defined by their structuring as spatial systems.	12

Environmental Change

Climatic Regions

Space and Population

How large is the population of the world, and what are the immediate prospects for growth?	See Fig. G-8.	20-20
Where are the four largest population agglomerations located?	In descending size, South Asia, East Asia, Europe, and Eastern North America.	19
What are the main features of the East Asian population cluster?	China's Huang and Chang / Yangzi river valleys; dominantly rural population.	19
What are the main features of the South Asian population cluster?	India's Ganges Lowland, Pakistan's Indus Valley, and all of Bangladesh in the Ganges-Brahmaputra Delta; heavily rural.	19
What are the main features of the European population cluster?	A central east-west population axis, oriented to industrial resources and highly urbanized.	19-20
What are the main features of the North American population cluster?	A pattern like the European cluster, dominated by metropolitan complexes along the northeastern U.S. seaboard.	20

Cultural Landscape

How can humans leave their imprint on the land surface?	In numerous ways, because people are agents of change; their structures and artifacts progressively transform natural into cultural landscapes.	21
What exactly is the *cultural landscape*?	According to Carl Sauer, the composite imprints and forms superimposed on the physical landscape by human activities.	27

Political Geography

| How many political units are there today? | In the year 2011, approximately 200 national states. | 23 |

| What is the *European state model*? | A state with a legally defined territory, inhabited by a population governed from a capital city by a representative government. | 24 |

Economic Geography

| What is the central concern of economic geography? | Studying the varied ways in which people earn a living, and how the goods and services they produce are spatially distributed and organized. | 28 |

| What are the four economic groups that countries are divided into? | High income, upper-middle income, lower-middle income, and low income. The global pattern of these countries is mapped in Fig. G-11. | 28-29 |

Patterns of Development

| Why is the world no longer simply divided into "developed" and "undeveloped" countries? | Many countries, despite their overall level of income, display cores of development that resemble rich societies, and peripheries of extreme poverty. | 31 |

| What are some of the obstacles that under-privileged areas face? | Beyond a disadvantageous position in the global economic system, less advantaged countries exhibit political instability, corrupt leadership, misdirected priorities, misused aid, and traditionalism. | 33-34 |

Geographic Realms

| What is a *geographic realm*? | A geographic realm is a large regional unit based on broad multiple criteria—reflecting physical, economic, political, urban, historical, and population as well as cultural geography. | 8-9; 35-37 |

What does the distribution of world geographic realms look like?	This framework, which underlies the organization of the rest of this book, are mapped in two levels of detail in Figs. G-3 and G-14; both maps should be studied carefully.	10-11; 36-37
What are the names of the 12 geographic realms?	Europe, Russia, North America, Middle America, South America, North Africa/Southwest Asia, Subsaharan Africa, South Asia, East Asia, South-east Asia, the Austral Realm, and the Pacific Realm.	35-37

Regional and Systematic Geography

What are the *systematic* subfields of geography?	The topical branches of the discipline that cut across the world realms, such as political or economic geography; the systematic subfields are diagrammed in Fig. G-15 on p. 38.	37-38

MAP EXERCISES

Map Comparison

1. Comparing Figs. G-5 and G-4, what generalizations can be made about the landforms that are located along tectonic plate boundaries, especially where plates are pushing against one another?

2. Comparing Figs. G-1 (pp. 2-3) and G-6 (p. 15), what generalizations can be made about the location of Pleistocene ice sheets in the Northern Hemisphere and the distribution of mountain ranges?

3. On Fig. G-8 (pp. 20-21), compare and contrast the *internal* patterns of the world's largest population agglomerations. Can some of the variations be accounted for by differences in environmental patterns as mapped in Fig. G-7?

4. Compare the advantages and disadvantages of generalizing world climate-region patterns (Fig. G-7 on pp. 16-17), with special emphasis on the mainland United States.

5. Carefully compare the map of world political units with the world population distribution map (Figs. G-11 and G-8, respectively). List five countries that are heavily populated but not large in area on the world political map, and five countries that are large in territorial size on the political map but modest in appearance in terms of population.

Map Construction (*Use outline maps at the end of this chapter*)

1. In order to examine and study the interrelationships among environmental systems, draw in the following for each of the world's continents:

    a. in *light red*, sketch in all major highland areas (use Fig. G-1 as a source, which is found on pp. 2-3 in the textbook)

    b. in *black* or *blue*, draw in all climate-zone boundaries from Fig. G-7 (pp. 16-17) and label each climate region with its appropriate Köppen-Geiger letter symbol.

2. Using Fig. G-8 (p. 20-21) as a guide, draw in and label each of the major world population agglomerations discussed on p. 19 in the text.

3. Using Fig. G-14 (pp. 36-37) as a guide, draw in the boundaries of all of the 12 world geographic realms and label each realm; *you should be able to do this by memory if you have learned Fig.G-14 first.*

PRACTICE EXAMINATION

Short-Answer Questions

Multiple Choice

1. The islands of the Caribbean Sea belong to which of the following realms:

    a) North America   b) Middle America   c) South America
    d) the Atlantic Realm   e) the Pacific Realm

2. The Mediterranean climate is classified under which of the following Köppen-Geiger letters:

    a) C   b) H   c) D   d) B   e) M

3. Which of the following realms does not contain a major world-scale population cluster?

    a) South America   b) North America   c) Europe
    d) South Asia   e) East Asia

4. Which is likeliest to exhibit a core area?

    a) transition zone   b) regional periphery   c) formal region
    d) uniform region   e) functional region

5. Which is not associated with a region's physiography?

    a) tectonic plate   b) climate pattern   c) cultural landscape
    d) effects of glaciation   e) dominant vegetation

True-False

1. The world's population is now over 8 billion in total size. **F**

2. Southeast Asia does not rank among the world's three largest population agglomerations. **T**

3. Interglacials are relatively warm spells that follow glaciations. **T**

4. Geographic realms almost never change over the course of a millennium. **F**

5. A city-hinterland relationship is an example of a functional region. **T**

Fill-Ins

1. The large country at the center of the realm called South Asia is _India_.

2. The _functional_ region is marked not by an internal sameness, but by the integration of a set of places and their interactions.

3. Desert and steppe climates both belong to the Köppen-Geiger category, included under the letter _B_.

4. The composite of the human imprint on the Earth's surface is called the _cultural landscape_

5. The two countries that constitute the Austral Realm are Australia and _New Zealand_

6. A region's _relative_ location refers to its spatial position with respect to surrounding regions.

Essay Questions

1. Identify and discuss the major geographic characteristics of your home region. Include, in your narrative, substantive answers to the following questions: Is it a formal or functional region, or does it contain aspects of both? How strong a role does urbanism play and, if prominent, are you located in the urban center itself or the hinterland? Are the boundaries of your region sharply defined or do they appear as broad zones of transition?

2. The cultural landscape of a region is one of its most important geographic components. Define and discuss this concept, taking into account the role of culture itself, its translation into a spatial expression on the land surface, and the significance of the special qualities that offset one cultural landscape from another.

3. The spatial distribution of population has been called the most essential of all geographic expressions because it represents the totality of human adjustments to the environment at that moment in time. Identify and discuss the global distribution of humankind, focusing on the four largest population agglomerations and the reasons why so many people have concentrated in those locales.

4. The *geographic realm* is the regional unit that forms the basis of the study of world geography in this book. Discuss what is meant by this regional term, using examples chosen from Fig. G-14.

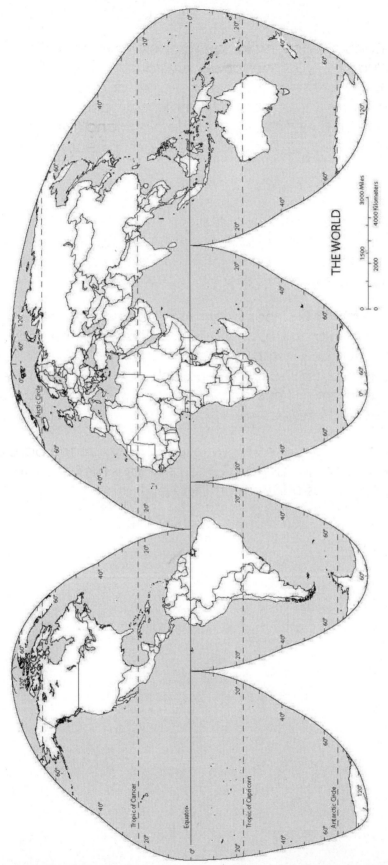

THE WORLD

0 1500 3000 Miles
0 2000 4000 Kilometers

Arctic Circle

Tropic of Cancer

Equator

Tropic of Capricorn

Antarctic Circle

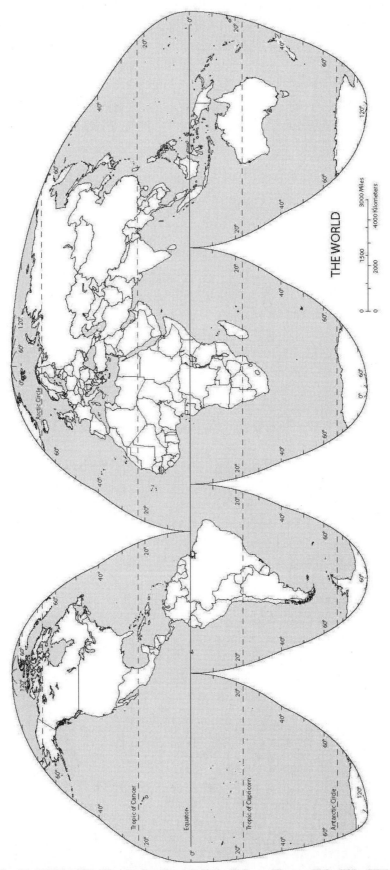

THE WORLD

3000 Miles
4000 Kilometers
1500
2000

Arctic Circle

Tropic of Cancer

Equator

Tropic of Capricorn

Antarctic Circle

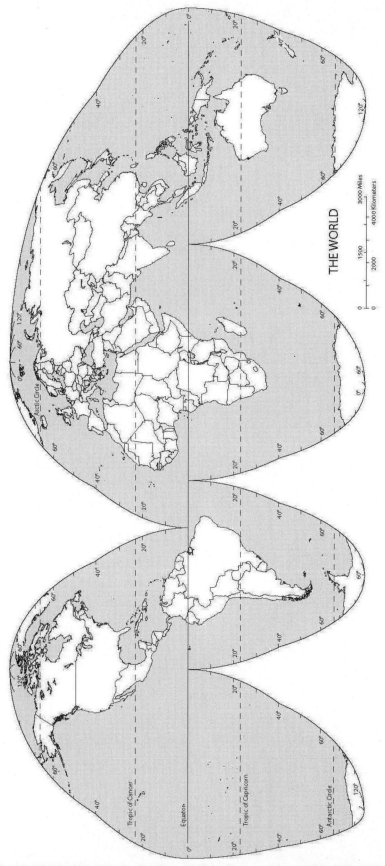

THE WORLD

0 1500 3000 Miles
0 2000 4000 Kilometers

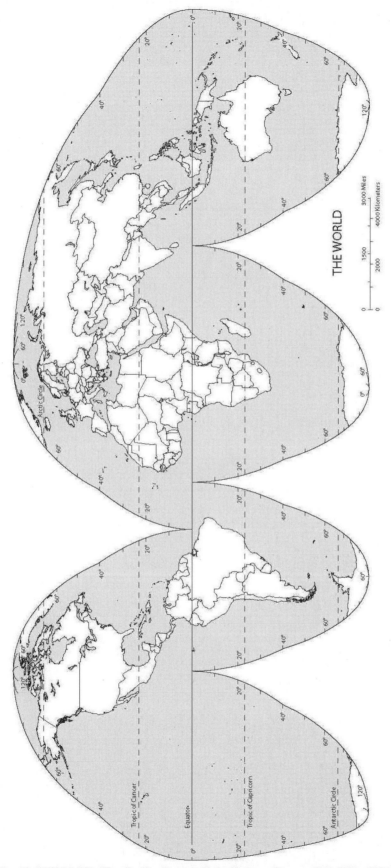

THE WORLD

3000 Miles
4000 Kilometers
1500
2000
0
0

Arctic Circle

Tropic of Cancer

Equator

Tropic of Capricorn

Antarctic Circle

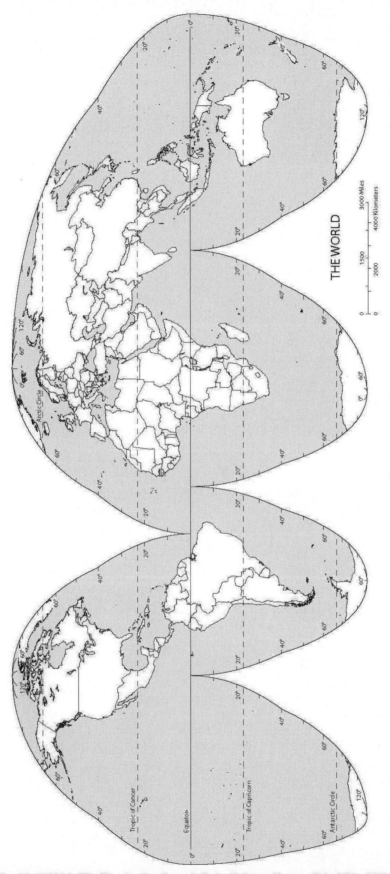

THE WORLD

0   1500   3000 Miles
0   2000   4000 Kilometers

CHAPTERS 1A and 1B
EUROPE

OBJECTIVES OF THIS CHAPTER PAIR

Chapters 1A and 1B, focus on the geography of Europe. Chapter 1A includes an introduction to Europe's advantageous geographic properties, its physical, historical, economic, political, and urban geography are discussed in their realm-wide contexts; the concluding sections treat European supranationalism and consider current trends as well as future prospects for greater unification. Chapter 1B is devoted to an examination of Europe's internal regions, proceeding through Europe's core and then the various components of its periphery.

Having learned the regional geography of Europe, you should be able to:

1. Identify Europe's remarkable geographic properties that propelled this modestly-sized corner of the world into international prominence and advanced economic development.

2. Understand Europe's broad physical geography, especially its subdivisioning into four major physiographic regions.

3. Appreciate the rich historical heritage of European society, whose modern foundation rests upon numerous economic and political revolutions.

4. Discern the current geographic dimensions of the realm, including the problems faced by aged industrial areas in a new postindustrial age.

5. Understand the intensifying trend toward greater European unification, the progress that has been made, and the challenges faced by the European Union in consolidating its gains and overcoming forces that would drive it apart.

6. Name the regional components of Europe, and the countries they contain.

7. Understand such basic politico-geographical concepts as nation-state, devolution, centripetal/centrifugal forces, irredentism, balkanization, and supranationalism.

8. Locate the major features of Europe on a map, including its countries, prominent physical regions, primary rivers, leading industrial areas, and largest urban centers.

CHAPTER 1A

Physiography (43)

The total physical geography of the terrestrial world.

Local functional specialization (46)

The production of a particular good or service as a dominant activity in a particular location.

Mercantilism (46)

Era of competitive accumulation of wealth, chiefly of gold and silver, after the discovery of new continents and riches.

Industrial Revolution (47)

A major turning point in a region's economic development whereby manufacturing becomes a leading growth activity, engendering further technological breakthroughs, capital investments, urbanization, and (eventually) a demographic transition.

Nation (50)

A group of tightly-knit people possessing bonds of language, ethnicity, religion, and other shared cultural attributes. Such homogeneity actually prevails within very few states.

Nation-state (50)

A political unit wherein the territorial state coincides with the area settled by a certain historical group of people, considering themselves to be a *nation*.

Spatial interaction (50)

Movement of any sort across geographic space, usually involving the contact of people in two or more places for the purposes of exchanging goods or services.

Complementarity (50)

Regional complementarity exists when two regions, through an exchange of commodities, can specifically satisfy each other's demands.

Transferability (50-51)

The capacity to move a good from one place to another at a bearable cost; the ease with which a commodity may be transported.

Metropolis (52)

A central city and its suburban ring.

Central Business District (CBD) (52)

The downtown of a large metropolitan area; the oldest part of an urban agglomeration containing the region's largest concentration of business, government, and shopping facilities and its wealthiest and most prestigious residences.

Centrifugal forces (54)

Disunifying or divisive forces that internally weaken a state.

Centripetal forces (54)

Forces that tie a state together, unifying and strengthening it.

Supranationalism (56)

A venture involving three or more national states—political, economic, or cultural cooperation to promote shared objectives. The European Union is such an organization.

European Union (55-table; 56-57)

Formerly called the Common Market or European Economic Community, it currently consists of 27 member countries: France, Italy, Germany, Belgium, the Netherlands, Luxembourg, the United Kingdom, Ireland, Denmark, Greece, Spain, Portugal, Austria, Finland, Sweden, Cyprus, the Czech Republic, Estonia, Hungary, Latvia, Lithuania, Malta, Poland, Slovakia, Slovenia, Bulgaria, and Romania.

Four Motors of Europe (58)

Rhône-Alpes (France), Baden-Württemberg (Germany), Catalonia (Spain), and Lombardy (Italy). Each is a high-technology-driven region marked by exceptional industrial vitality and economic success not only within Europe but on the global scene as well.

Devolution (60)

The process whereby regions within a state demand and gain political strength and growing autonomy at the expense of the central government.

CHAPTER 1B

Microstates (69)

Tiny entities that do not have all of the attributes of "complete" states, such as Monaco, San Marino, Andorra, and Liechtenstein.

Site (72)

The *internal* locational attributes of a place, including its local spatial organization and physical setting.

Situation (72)

The *external* locational attributes of a place; its *relative location* or position with reference to other non-local places.

Conurbation (76)

General term used to identify large multi-metropolitan complexes formed by the coalescence of two or more major urban areas. Randstad Holland in the Netherlands is a leading European example.

Landlocked location (77)

An interior country that is surrounded by land, such as Switzerland and Austria.

Hegemony (78)

Political dominance.

Shatter belt (88)

A zone of chronic political splintering and fracturing—Eastern Europe and Southeast Asia are classic examples.

Balkanization (88)

The fragmentation of a region into smaller, often hostile political units.

Ethnic cleansing (89)

The forcible ouster of entire populations from their homelands by a stronger power who seeks to conquer their territory.

Break-of-bulk point (92)

A location along a transport route where goods must be transferred from one carrier to another. In a port, the cargoes of oceangoing ships are unloaded and put on trains, trucks, or perhaps smaller boats for inland distribution.

Entrepôt (92)

A place, usually a port city, where goods are imported, stored, and transshipped. Thus an entrepôt is a *break-of-bulk point*.

Exclave (93)

A bounded (non-island) piece of territory that is part of a particular state but lies separated from it by the territory of another state. Alaska is an exclave of the United States.

Irredentism (97)

A policy of cultural extension and potential political expansion by a country aimed at a group of its nationals living in a neighboring country.

SELF-TESTING QUESTIONS

Cover the right side of the page with a sheet of paper. Uncover each line after you have attempted to answer the question in the left column. If necessary, refer to the textbook page(s) listed at the right.

Question	Answer	Page
CHAPTER 1A		
Europe's Characteristics		
Where is Europe's eastern boundary?	Some scholars insist it is the Ural Mountains, while others argue that all of western Russia is a transition zone; this text uses the boundary between Eastern Europe and Russia.	42-43
What types of raw materials spawned Europe's development?	Early on rich soils, good fishing waters, domesticatable animals, and plentiful wood for building. Later, mineral fuels and ores made industrialization possible.	43
What are Europe's chief geographic resources?	Its unmatched advantages of scale and proximity; a diverse environment and cultural mosaic, facilitating innovation and easy contact with the rest of the world.	43-44
Physical Landscapes		
What are the major characteristics of the Central Uplands?	Forms the heart of Europe; a resource-laden belt where Industrial Revolutions and cities emerged in the 19th century.	43
What are the major characteristics of the Alpine Mountains?	The Alps and their outliers (the Pyrenees, Dinaric Alps, and Carpathians) forming a high-mountain backbone.	43-44
What are the major characteristics of the Western Uplands?	Rugged, older highlands often located along stormy oceanic fringes; represent earlier geological mountain building than the Alpine system.	44

What are the major characteristics of the North European Lowland?	Densest populations of the realm are found here; historic route of human contact and migration, yet much internal variation in generally low-lying terrain; a-lso contains much of Europe's most productive agriculture and a multitude of navigable rivers.	44

The Revolutions

What was the *Industrial Revolution*?	The rapid growth of mass manufacturing through mechanization that triggered far-reaching social and economic change, demographic transition, and mass urbanization.	47
Has the diversity of ancestries in Europe been an asset or a liability to the realm?	Both–it has generated interaction and exchange, but has also caused conflict and war.	48-50
How did the political revolution after 1780 forever change Europe?	Monarchies gave way to republics, democracy flourished, and nationalism became the dominant political force.	48-49
What is a *nation*?	A group of tightly-knit people possessing bonds of language, ethnicity, religion, and other shared cultural attributes. Such homogeneity actually prevails within very few states.	50
What is a *nation-state*?	A political unit wherein the territorial state coincides with the area settled by a certain historical group of people, considering themselves to be a nation.	50

Europe Today

Is Europe an overall homogeneous regional unit?	To a surprising degree it is not—language, religion, and race are marked by much diversity; yet this diversity, through various sub-regional complementarities and cooperation, has been overcome to form a well-unified realm.	50-52

| Define the two key principles of interaction | *Complementarity*, which exists when two places, through an exchange of goods or services, can specifically satisfy each other's demands; *transferability*, the capacity to move an item from one place to another at a bearable cost. | 50-52 |

European Urbanization

Where does Europe rank among the urbanized realms?	Nearly 75% of Europe's population lives in cities and towns. See also Appendix B.	52
What is a metropolis?	A central city plus its suburban ring.	52
What are *centripetal* and *centrifugal* forces?	*Centripetal forces* unify a state; *centrifugal forces* are divisive or disunifying.	54

European Devolution and Unification

What are Benelux and the EEC? Name the countries of the European Union.	Benelux (1944) involved the economic association of Belgium, the Netherlands, and Luxembourg; EEC (1958) was the European Economic Community, originally comprised of France, (then West) Germany, Italy, and the Benelux nations. The countries of the European Union (EU) today are France, United Kingdom, Germany, Italy, Greece, Denmark, Ireland, Spain, Portugal, the 3 Benelux countries, Finland, Austria, Sweden, Cyprus, the Czech Republic, Estonia, Hungary, Latvia, Lithuania, Malta, Poland, Slovakia, Slovenia, Romania, Bulgaria.	55-table; 56-57
What are the Four Motors of Europe?	1) Rhône-Alpes region, 2) Lombardy, 3) Catalonia, 4) Baden-Württemberg. These subnational regions are hubs of economic power and influence which have developed direct linkages with one another, bypassing the capitals and governments of their countries.	58
What is devolution?	The process whereby regions within a state demand and gain political strength and growing autonomy at the expense of the central government.	60

CHAPTER 1B

Western Europe

Which countries constitute this region?	France, Germany, Belgium, the Netherlands, Luxembourg, Switzerland, Austria, and Liechtenstein. See Fig. 1B-2, p. 70.	70
How long was Germany partitioned into East and West?	From the end of World War II (1945) until the reunification of Germany in 1990.	69-71
What geographical features contributed to West Germany's post-World War II "economic miracle?"	Large and varied amounts of raw materials, port access, a location neighboring all other prosperous European nations, efficient surface transport systems, and security protection from NATO all helped West Germany to prosper.	69-70
What is the difference between site and situation?	Site refers to the internal, local physical and other attributes of a place; situation is the geographical position of a place or its external locational qualities.	72
Define conurbation.	General term denoting a large multi-metropolitan complex formed by the coalescence of two or more major urban areas; the Randstad in the western Netherlands is an outstanding example.	76
What is the Randstad?	The Dutch "ring-city" conurbation linking Amsterdam, The Hague, and Rotterdam.	76
What are Switzerland's two leading cities?	Zurich and Geneva.	77
What city and river are synonymous with Austria?	Vienna and the Danube.	77
In what way does Vienna represent an outpost?	It is the Mainland's Core easternmost city of Western Europe, but its relative location has become more centralized with the eastward shift of the EU border.	77

The British Isles

What are the political components of this region?	Britain—containing England, Wales, and Scotland—and Ireland, which contains the Irish Republic (Eire) and Northern Ireland; all but Eire form the United Kingdom. See Fig. 1B-6, p. 78	77-78
What are the United Kingdom's four sub-regions, and how are they characterized?	Dominant England, individualistic Scotland, intractable Wales, and embattled Northern Ireland.	79-80
What are the economic prospects for Southern England?	High-technology and service industries, such as financial, engineering, communications, and energy-related activities are flourishing in the South.	79
Which two nearby cities are the industrial and cultural foci of Scotland?	Respectively, Glasgow and Edinburgh.	79
Who are the contestants for power in Northern Ireland?	The Protestants who now barely constitute 50 percent of the residents, vs. the now growing 46 percent who are Roman Catholic; the British are trying to prevent a civil war.	80

Southern Europe

Which countries constitute this region?	Italy, Spain, Portugal, Greece, Cyprus, and Malta.	81-82
What generalization can be made about the population distributions of this region's countries?	They are decidedly peripheral with large concentrations in productive coastal and riverine lowlands, with more growth of major industrial centers in northern Italy and Spain.	81-82
How do the levels of urbanization and raw materials of this region compare to those to the north?	Urbanization lags considerably, reflecting agricultural bases of the preindustrial era and higher population growth rates. Raw materials are lacking and must be imported from the European core.	81-82

Which country is Mediterranean Europe's most populous and best connected to the European core?	Italy.	82
Which vital subregion of Italy contains nearly half its population and today functions as its economic core?	The Po River basin of the north, which focuses on Lombardy, containing the largest Italian city (Milan), and also belongs to Europe's core area (Fig. 1B-8, p. 82).	82-83
What is meant by the term Mezzogiorno?	This is the name for Italy's poverty-stricken and stagnant agricultural south (southeast of the Ancona Line).	82-83
Which countries are located on the Iberian Peninsula?	Spain and Portugal.	83-84
Where is Spain's major manufacturing zone located?	In Catalonia, in the northeast of the country.	84
How has EU membership affected Greece economically?	Its economy was booming, and Greece had been described as a locomotive for the Balkans. As of 2011, the country was mired in debt as EU subsidies were diverted to Bulgaria and Romania.	86
How does the northern area of Cyprus differ from its southern counterpart?	The Turkish-proclaimed Turkish Republic of Northern Cyprus is north of the UN "Green Line," contains about 40 percent of Cyprus' territory, and has about 100,000 inhabitants. The south has 60 percent of the territory, about 900,000 inhabitants, and a more prosperous economy with strong links to Europe.	87-88

Nordic Europe (Norden)

Which countries constitute this region?	Norway, Denmark, Sweden, Iceland, Finland, and Estonia. See Fig 1B-12	90-91

Why is southern Sweden much more densely populated than northern Sweden?	In southern Sweden, relief is lower and more manageable, soils are better, and the climate is milder.	90
How have the seas favored Norway in recent years?	Huge deposits of oil and natural gas in the Norwegian sector of the North Sea yield both fuel supplies and heightened export income.	91-92
What activities form the base of Finland's economy?	Wood and wood products are the major exports. The manufacture of machinery, growth of staple crops, and telecommunications also play important roles.	93
Why is Estonia considered to be part of Northern, rather than Eastern Europe?	Estonia shares close linguistic, ethnic, and historical-geographical ties with Finland.	93

Eastern Europe

What is meant by the term balkanization? Why is this region so often called a shatter belt?	The fragmentation of a region into smaller, often hostile political units. Because of the numerous cultures that have collided here, the resultant shattering of various cultural and political units has produced an especially fragmented ethnic map.	88
What is ethnic cleansing?	The forcible removal of entire populations from their homelands by a stronger power determined to take their territory.	89
What is the largest new country in the region?	Serbia, with 7.2 million inhabitants.	89
What religion dominates Albania?	The country is more than 75% Islamic.	90
Why was Ukraine an economic-geographic cornerstone of the former Soviet Union?	Ukraine is rich in industrial raw materials and contained the most productive environment (soils; climate) for large-scale agriculture in the U.S.S.R.; Ukrainian mineral and energy deposits gave rise to the Donbas, until the 1990s one of Eurasia's leading regions of heavy manufacturing.	98

MAP EXERCISES

Map Comparison

1. Compare the map of Europe's relative world location (Fig. 1A-4, p. 46) with the world political map (Fig. G-11, p. 22); discuss the differences between the relative and absolute location of Europe, and how different map projections can convey very different perspectives.

2. The boundary of Europe's core area is carefully drawn in Fig. 1B-2 (p. 70). Briefly trace that boundary through each country it lies in and, by referring to other maps and appropriate text, justify why the line follows the route it does.

3. The languages of Europe perform a variable function in the realm's political units, sometimes binding people together and sometimes dividing them further. By reviewing Fig. 1A-7 (pp. 51-52) and appropriate text, comment on the status of language as a political force in each of the following countries: United Kingdom, Belgium, Switzerland, Poland, Finland, and Romania.

Map Construction (Use outline maps at the end of this chapter)

1. In order to familiarize yourself with Europe's physical geography, place the following on the first of the outline maps:

   a. Rivers: Thames, Seine, Loire, Rhône, Garonne, Meuse (Maas), Rhine, Elbe, Danube, Po, Ebro, Tagus, Guadalquivir, Vistula, Oder, Neisse, Tisza, Sava, Dniester, Dnieper

   b. Water bodies: Mediterranean Sea, Baltic Sea, North Sea, Irish Sea, English Channel, Strait of Gibraltar, Tyrrhenian Sea, Adriatic Sea, Ionian Sea, Black Sea, Sea of Azov, Bay of Biscay, Gulf of Finland, Gulf of Bothnia

   c. Land bodies: Iceland, Ireland, Sicily, Cyprus, Malta, Corsica, Sardinia, Balearic Islands, Shetland Islands, Scandinavian Peninsula, Jutland Peninsula, Iberian Peninsula, Peloponnesus, Crimea Peninsula

   d. Mountains: Alps, Pyrenees, Appennines, Dinaric Alps, Pennines, Scottish Highlands, Dolomites, Tatras, Ore Mountains (Erzgebirge), Sudeten Mountains, Jura, Cantabrians, Carpathians, Balkans, Transylvanian Alps, Rhodope Mountains, Pindus Mountains

   e. Other landforms: Meseta, Po Plain, Massif Central, North European Lowland, Bohemian Basin

43

2. On the second map, political-cultural geographic information should be entered as follows:

   a. Label each country that is listed in Table G-1 table under Europe.

   b. For each of those countries locate and label the capital city with the symbol *.

   c. Reproduce the language map using light pencil coloring (Fig. 1A-7, p. 39).

3. On the third outline map, economic-urban information should be entered as follows:

   a. Europe's regions: Using a thick line, draw the boundaries of Europe's Core

   b. Supranational affiliations: Color the countries of the European Union green, and countries that do not belong to the European Union yellow.

   c. Cities (locate and label with the symbol ●): London, Birmingham, Manchester, Liverpool, Newcastle, York, Leeds, Edinburgh, Glasgow, Dublin, Belfast, Paris, Marseille, Lyon, Bordeaux, Toulouse, Strasbourg, Antwerp, Brussels, Luxembourg, Rotterdam, Amsterdam, Hamburg, Cologne, Essen, Düsseldorf, Frankfurt, Berlin, Stuttgart, Munich, Zürich, Geneva, Vienna, Graz, Copenhagen, Oslo, Bergen, Göteborg, Stockholm, Helsinki, Milan, Turin, Genoa, Venice, Florence, Rome, Naples, Palermo, Madrid, Lisbon, Seville, Barcelona, Athens, Istanbul, Warsaw, Krakow, Katowice, Prague, Bratislava, Budapest, Bucharest, Sofia, Trieste, Belgrade, Ljubljana, Zagreb, Sarajevo, Pristina, Skopje, Tirane, Chisinau, Odesa, Kiev (Kyyiv), Lviv, Dnipropetrovsk, Donetsk, Kharkiv, Mensk (Minsk), Kaliningrad, Vilnius, Riga, and Tallinn

   d. Economic regions (identify with circled letter):

      A -the Ruhr
      B -Paris Basin
      C -Rhône-Alpes Region
      D -Baden-Württemberg
      E -Silesia
      F -Randstad Holland
      G -Catalonia
      H -Lombardy
      I - the Donbas

Short-Answer Questions

Multiple-Choice

1. Which of the following mountain ranges is regarded by a number of geographers as Europe's eastern boundary?

    a) Pennines   b) Appennines c) Pyrenees   d) Urals   e) Alps

2. Which of the following cities is located in Italy's sector of the European core area?

    a) Milan   b) Catalonia   c) Barcelona   d) Naples   e) Rome

3. What is Spain's leading industrial area?

    a) Madrid   b) Lisbon   c) Andalusia   d) Catalonia e) Lombardy

4. In which of the following did the Industrial Revolution begin?

    a) Russia   b) Germany   c) Scotland d) England e) France

5. Which of the following countries is not a member of the European Union?

    a) Poland   b) Serbia   c) Cyprus d) Bulgaria e) Sweden

6. Which of the following political entities was not a part of former (pre-1990) Yugoslavia?

    a) Albania   b) Montenegro   c) Croatia d) Slovenia e) Macedonia

True-False

1. The Republic of Ireland (Eire) is not a part of the state called the United Kingdom. T

2. Although it pursued the acquisition of territory and precious metals, mercantilism was not concerned with actively spreading Christianity throughout the New World. F

3. The spatial interaction principle of transferability refers to the capacity to move a good at a bearable cost. F

A 4. All of the British Isles lie outside of the Central European Upland region.

4 5. The major internal regions of Italy are called Autonomous Communities.

Fill-Ins

1. The chief territorial occupant of the Iberian Peninsula is the country of
   Spain.

2. A politico-geographical force that operates to unify a country is known as a
   Centripetal force.

3. _Devolution_ is the term used to describe the situation in which the
   regions or peoples within a state gain political strength and sometimes
   autonomy at the expense of the center.

4. Greenland is politically affiliated with the northern European country of
   Denmark.

5. The eastern Mediterranean island of _Cyprus_ is divided between
   Greece and Turkey.

Matching Question on European Countries

I	1.	Port of Rotterdam	A.	Bosnia
E	2.	Lyon high-tech region	B.	Ukraine
M	3.	Scottish devolution threat	C.	Czech Republic
O	4.	Home of "White" Russians	D.	Denmark
J	5.	Scandinavian Peninsula	E.	France
G	6.	Appennine Mountains	F.	Germany
B	7.	Crimea Peninsula	G.	Hungary
L	8.	Centered on Belgrade	H.	Italy
N	9.	Catalonian autonomy drive	I.	The Netherlands
D	10.	Jutland Peninsula	J.	Norway
K	11.	Southwestern Iberia	K.	Portugal
G	12.	Home of the Magyars	L.	Serbia
A	13.	Large Muslim population	M.	United Kingdom
F	14.	Land of Länder	N.	Spain
C	15.	Bohemian Basin	O.	Belarus

Essay Questions

1. One of the major turning points in European (and world) history was the achievement of the Industrial Revolution. Discuss the sequence of events that marked this economic transformation, highlight its geographic expressions, and show how it helped shape the current map of European manufacturing activity.

2. One of the major geographic qualities of Europe is its strong regional differentia-tion, which has given rise to a high degree of areal functional specialization affording multiple exchange opportunities. Provide three examples of such spatial complementarity and discuss the interaction that each has produced over the past century.

3. The partitioning of Germany was one of Europe's major politico-geographical events of the 20th century—a situation that ended in 1990 with the reunification of West and East Germany. Discuss the resource bases and economic-geographic infrastructures of the Länder of former East and West Germany, and show how the reunification has produced financial and social strain.

4. Eastern Europe, the arena for cultural collisions and conflicts for centuries, is a classic example of a shatter belt. Discuss the concept of the shatter belt and the balkanization it has produced here, and give several examples of persistent cultural fragmentation (and turmoil) that continue here today.

5. Discuss the current state of unification in the European realm under the banner of the EU. Using the table on page 55, review the future possibilities of membership for each country that lies outside of the EU today.

ICELAND

NORWEGIAN SEA

Shetland Islands

ATLANTIC OCEAN

Scottish Highlands

Pennines

IRELAND

IRISH SEA

CELTIC SEA

Thames R.

ENGLISH CHANNEL

Seine R.

Loire R.

BAY OF BISCAY

Cantabrian Mts.

Iberian Peninsula

Tagus R.

Guadalquivir R.

STRAIGHT OF GIBRALTAR

Garonne R.

PYRENEES

Ebro R.

Balearic Islands

MEDITERRANEAN SEA

Corsica

Sardinia

TYRRHENIAN SEA

SICILY

Malta

NORTH SEA

JUTLAND PENINSULA

Meuse R.

Rhine R.

Jura Mts.

Rhone R.

Po R.

ALPS

APENNINES

Scandinavian Peninsula

GULF OF BOTHNIA

GULF OF FINLAND

BALTIC SEA

Elbe R.

Neisse R.

Oder R.

Sudeten Mts.

Vistula R.

Tatra Mts.

CARPATHIAN MTS.

Tisza R.

Ore Mountains

TRANSYLVANIAN ALPS

Danube R.

Dolomites

DINARIC ALPS

Sava R.

ADRIATIC SEA

BALKAN MTS.

Rhodope Mts.

Pindus

IONIAN SEA

Peloponnesus

Dnieper R.

Dniester R.

CRIMEA PENINSULA

SEA OF AZOV

BLACK SEA

CYPRUS

EUROPE

	200	400	600 Kilometers
0			
0	100	200	300 Miles

Arctic Circle

30° 70° 20° 10° 0° 10° 20° 30° 40° 50° 70°

60°

50°

40°

30°

0° 10° 20° 30°

Key (capital cities):

1 - Tirane, Albania
2 - Brussels, Belgium
3 - Prague, Czech Republic
4 - Bratislava, Slovakia
5 - Ljubljana, Slovenia

6 Budapest, Hungary
7 Zagreb, Croatia
8 Sarajevo, Bosnia
9 Belgrade, Serbia
10 Podgorica, Montenegro

11 Pristina, Kosovo
12 Skopje, Macedonia
13 Vilnius, Lithuania
14 Luxembourg, Luxembourg
15 Innsbruck, Liech

Copyright © 2010 John Wiley & Sons, Inc. From Geography: Realms, Regions, and Concepts, 14e by deBlij and Muller

EUROPE

0 200 400 600 Kilometers

0 100 200 300 Miles

Arctic Circle

EUROPE

0   200   400   600 Kilometers
0 100   200   300 Miles

Arctic Circle

EUROPE

| 0 | | 200 | | 400 | | 600 Kilometers |
| 0 | 100 | | 200 | | 300 Miles | |

Arctic Circle

CHAPTERS 2A and 2B
RUSSIA

OBJECTIVES OF THESE CHAPTERS

Chapters 2A and 2B cover Russia, former heart of the Soviet Union, a troubled realm, and a society and economy undergoing profound upheaval and transformation. Russia comprises the world's largest political unit in areal size and contains every climate type except the wet tropical category. After the changes engendered by the 1991 collapse of the U.S.S.R. are briefly outlined. Russia's physiographic, economic, ethnic, and geopolitical frameworks set the stage for a survey of the realm's historical geography and modern regions. Chapter 2B also contains an overview of the conflict-plagued Transcaucasian Transition Zone in the section that covers Russian Peripheries.

Having learned the regional geography of Russia, you should be able to:

1. Understand the overall climatic pattern of this realm.

2. Grasp the essential ingredients of Russia's complex historical evolution, including the contributions of major groups that effected lasting change.

3. Understand the significance of the Soviet period and its legacy for Russia's current circumstances.

4. Understand Russia's cultural-geographic mosaic and its relationship to the churning matrix of internal republics.

5. Appreciate the ongoing spatial reorganization of Russia's economy necessitated by the detachment of its 14 neighboring former Soviet Socialist Republics, especially Ukraine and Kazakhstan.

6. Understand Russia's international political boundary problems, particularly in the eastern reaches of the country.

7. Map and describe the leading functions of the realm's four major regions.

8. Appreciate the ongoing ethnic and territorial conflict in the shatter belt of Transcaucasia.

9. Locate the major physical, cultural, and economic-spatial features of the realm on an outline map.

Chapter 2A

Continentality (108)

Refers to degree of inland location away from the moderating effects of the oceans on the climates of adjacent landmasses. The most interior locations, such as eastern Russia, experience both moisture deficits and very large annual temperature ranges.

Climate (109)

A term used to convey a generalization or synthesis of all the recorded weather observations over time at a certain place or in a given area; it represents an average of all the weather that occurs there.

Weather (109)

The state of the atmosphere at a location at any given moment. Recorded in terms of temperature, percentage of humidity, amount of precipitation, wind speed and direc-tion, and the like.

Tundra (109)

Treeless plain along the Arctic shore, containing mosses, lichens, and some grasses.

Taiga (109)

Mostly coniferous forests that cover large parts of northern Russia and Canada; located south of where the tundra ends.

Permafrost (109)

Permanently frozen ground in extreme northern Russia and Canada.

Global Climate Change (109-110)

Changes made to climate that occur naturally, often in cycles, although in more recent uses of the term it refers to human-induced changes (such as increased carbon dioxide in the atmosphere) that are accelerating natural climate change.

Forward capital (113)

Capital city positioned in actually or potentially contested territory, usually near an international border; it confirms the state's determination to maintain its presence in the region in contention. In the case of St. Petersburg, it demonstrated Russia's commitment to play a major role in the Baltic theater as well as its receptivity to European cultural and economic innovations.

Imperialism (115)

The drive toward the creation and expansion of a colonial empire and, once established, its perpetuation.

Russification (117)

Demographic resettlement policies pursued by the central planners of the Soviet Empire, whereby ethnic Russians were encouraged to emigrate to non-Russian republics of the U.S.S.R.

Federation (117)

A political framework wherein a central government represents the various subnational entities within a nation-state, where they have common interests–defense, foreign affairs, and the like–yet allows these various entities to retain their own laws, policies, and customs in certain spheres.

Soviet Socialist Republic (116-117)

One of the former 15 S.S.R.s that constituted the Soviet Union, each corresponding broadly to one of the country's major nationalities. The largest was the Russian Soviet Federative Socialist Republic (R.S.F.S.R.). Figure 2A-7on p. 89 maps the S.S.R.s within the now-defunct Soviet Empire.

Collectivization (117)

The reorganization of a country's agriculture under communism that involves the expropriation of private holdings and their incorporation into relatively large-scale units, which are farmed and administered cooperatively by those who live there.

Command Economy (117)

Involves highly centralized control of the national planning process, a hallmark of communist economic systems. Soviet central planners mainly pursued a grand political design in assigning production to particular places; their frequent disregard of economic geography contributed to the eventual collapse of the U.S.S.R.

Near Abroad (123)

The 14 former Soviet republics that, in combination with the dominant Russian Republic, constituted the USSR. Since the 1991 breakup of the Soviet Union, Russia has asserted a sphere of influence in these now-independent countries, based on its proclaimed right to protect the interests of ethnic Russians who were settled there in substantial numbers during Soviet times. These 14 countries include Armenia, Azerbaijan, Belarus, Estonia, Georgia, Kazakhstan, Kyrgyzstan, Latvia, Lithuania, Moldova, Tajikistan, Turkmenistan, Ukraine, and Uzbekistan.

Unitary state system (130)

Nation-state that has a centralized government and administration that exercises power equally over all parts of the state.

Distance decay (127)

The various degenerative effects of distance on human spatial structures and interactions.

Population implosion (130)

A decrease in national population. Russia is predicted to have only about 110 million people in 2050.

Core area (132)

The heartland of a state. The largest population cluster, the most productive region, the area with greatest centrality and accessibility.

Povolzhye (134-135)

Russian name for the basin of the middle and lower Volga River.

Kuzbas (136)

Contraction of "Kuznetsk Basin," the leading industrial zone of the Eastern Frontier region based on coal, iron ore, and good long-distance transport connections.

Transcaucasia (141)

The region constituted by Georgia, Armenia, and Azerbaijan; a mini-shatter-belt known as the "Balkans of Asia;" a historical and current region of ethnic and territorial conflict.

Cover the right side of the page with a sheet of paper. Uncover each line after you have attempted to answer the question in the left column. If necessary, refer to textbook page(s) listed at the right.

Question	Answer	Page

Chapter 2A

Climatology

Question	Answer	Page
Where is most of Russia's population concentrated?	In "European" Russia, west of the Ural Mountains.	109
What is the difference between weather and climate?	Weather is the state of the atmosphere at any given moment; climate is the average weather over a long period of time.	109
Why is farming difficult in Russia?	Even in its most favorable areas, temperature extremes, variable and undependable rainfall, and short growing seasons make agriculture a challenge for Russians.	109

Physiography

Question	Answer	Page
Where is the Russian Plain located?	Across all of "European" Russia as far east as the Urals; the Moscow Basin lies at its heart.	106
What is considered to be the world's largest unbroken lowland?	The West Siberian Plain.	107
What is the major inland water body of the Eastern Highlands?	Lake Baykal, the world's deepest.	107

Chapter 2B

Russian History

When did Moscow emerge as the center of the Russian state?	In the 16th century, when Czar Ivan the Terrible began to expel the Tartar conquerors.	113
How far did the Russians penetrate into North America?	They reached across Alaska, the western coast of Canada, and as far south as San Francisco Bay; the U.S. later purchased Alaska from Russia in 1867.	115
What were the contributions of Czar Peter the Great?	Peter opened up the Baltic frontier, built the capital of St. Petersburg, and developed ties with Europe; Catherine pursued the southern frontier, obtaining warm-water ports on the Black Sea.	113-115

Soviet Legacy

When was the Soviet Union formed?	Between 1917 and 1924, as a result of the Bolshevik Revolution that overthrew the czarist monarchy.	116
Which was the largest of the former S.S.R.s?	The Russian republic, officially known as the Russian Soviet Federative Socialist Republic (R.S.F.S.R.). See Fig. 2A-7, p. 116.	116-117
What was the overall result of past Soviet attempts at population planning?	Minority peoples were generally moved eastward and replaced with Russians. This was termed Russification.	117
Was the former Soviet Union truly a federation?	In theory, yes-any republic wishing to leave the federation supposedly could have. However, in reality, Moscow maintained total control over the republics.	117
What were the two main objectives of Soviet economic planners?	To speed industrialization and collectivize agriculture. Soviet planners hoped that efficient farming would free up the labor force necessary for industry.	117

Russia Today

Russia's Internal Republics

Regions of the Russian Realm

Transcaucasia

Where is Transcaucasia?	It lies between the Black Sea to the west and the Caspian Sea to the east. See Fig. 2B-6, p. 140.	141
Why is Transcaucasia called the Balkans of Asia?	It is a jigsaw of languages, religions, and ethnic groups–an historical battleground for Christians and Muslims, Russians and Turks, Armenians and Persians.	141
What are the three political entities of Transcaucasia?	Georgia, Armenia, and Azerbaijan, all former S.S.R.s of the Soviet Union.	141

Armenia

Where is Armenia's exclave?	Nagorno-Karabakh, inside neighboring Muslim Azerbaijan.	141-142

Georgia

Name two minority-based autonomous entities in this country.	Abkhazia and South Ossetia.	142
Which centrifugal forces have come to the forefront since the fall of the U.S.S.R.?	Factional fighting destroyed Georgia's first elected government in 1991; conflict continues in South Ossetia as well as in Abkhazia.	142

Azerbaijan

What are the republic's ties to its southern neighbor, Iran?	The Shi'ite Muslim Azeris of Azerbaijan have much in common with their ethnic brethren in Northern Iran; irredentism may be a threat in the future.	142
What future political problems does Azerbaijan face?	Fighting over Nagorno-Karabakh with the Armenians, and possibilities of Iranian and Turkish involvement in the dispute.	142

MAP EXERCISES

Map Comparison

1. Russian climates represent one pattern of regional distribution for a large landmass lying between approximately 30°N and 70°N. Compare this climate pattern to the one exhibited by North America (see Intro chapter) and note the similarities and differences.

2. Russia contains a sizeable non-Russian population, as the pre-1992 map (Fig. 2A-7, p. 111) reveals. Compare this map with the distribution of Russian ethnic groups (Fig. 2A-8, p. 118) and further compare both maps to Russia's population distribution (Fig. 2A-4, p. 110) noting those areas where the Russians are most under-represented (this should give you an insight into the internal spatial patterns of the 21 republics that are mapped in Fig. 2B-2, p. 128-129.

3. How do the Russian oil and gas regions (Fig. 2A-5, p. 111) and railway network (Figs. 2A-9, pp. 120-121 & 2B-5, p. 140) compare to the national population distribution (Fig. G-11, p. 22)?

4. Compare Russia's energy-producing (Fig. 2B-4) and manufacturing (Fig. 2B-3, p. 133) regions with appropriate maps in Chapter 1 (covering Ukraine, Moldova, Belarus, Lithuania, Latvia, and Estonia), Chapter 6 (covering the republics of Turkestan: Kazakhstan, Uzbekistan, Turkmenistan, Tajikistan, and Kyrgyzstan), and the section of Chapter 2 covering Transcaucasia (Georgia, Armenia, and Azerbaijan) that span the 14 S.S.R.s detached from Russia after the 1991 disintegration of the Soviet Union. What are Russia's most important losses? How might these new deficiencies be made up from both inside and outside Russia?

Map Construction

(Use outline maps at the end of this chapter)

1. In order to familiarize yourself with Russian physical geography, place the following on the first outline map:

   a. *Rivers:* Don, Dnieper, Volga, Dvina, Ural, Kama, Ob, Irtysh, Yenisey, Angara, Lena, Aldan, Kolyma, Amur, Ussuri, Tunguska

   b. *Water bodies:* Caspian Sea, Black Sea, Sea of Azov, Lake Baykal, Sea of Okhotsk, Bering Sea, Barents Sea, White Sea, Baltic Sea

   c. *Land bodies:* Kola Peninsula, Novaya Zemlya, Kamchatka Peninsula, Kurile Islands, Sakhalin, Moscow Basin, Crimea Peninsula, West Siberian Plain, Yakutsk Basin

   d. *Mountain Ranges:* Ural Mountains, Caucasus Mountains, Central Asian Ranges, Verkhoyansk Mountains, Sikhote Alin Mountains

   ✱ *label countries in the Caucasus*

2. On the second map, cultural information should be entered as follows:

   a. Reproduce the ethnic map (Fig. 2A-8, p. 118) in general form using an appropriate set of area symbols (preferably color pencils).

3. On the third outline map, economic-urban information should be entered as follows:

   a. *Cities* (locate and label with the symbol ●): Moscow, St. Petersburg, Rostov, Voronezh, Kursk, Tula, Nizhniy Novgorod, Ivanovo, Yaroslavl, Murmansk, Arkhangelsk, Perm, Kazan, Ufa, Samara, Saratov, Volgograd, Astrakhan, Groznyy, Nizhniy Tagil, Yekaterinburg, Chelyabinsk, Magnitogorsk, Qaraghandy, Astana, Omsk, Novosibirsk, Krasnoyarsk, Bratsk, Irkutsk, Yakutsk, Khabarovsk, Vladivostok, Nakhodka, Komsomolsk, Magadan, Tbilisi, Sokhumi, Batumi, Kutaisi, Yerevan, Gyumri, Baki (Baku)

   b. *Economic regions* (identify with circled letter):

   A - Central Industrial Region
   B - *Povolzhye*
   C - Urals Industrial Region
   D - *Kuzbas*
   E - Far East Region
   F - Transcaucasia

PRACTICE EXAMINATION

Short-Answer Questions

Multiple-Choice

1. What was the capital of the Soviet Union in the early years following the 1917 Bolshevik Revolution?

    a) Moscow   b) Leningrad   c) Petrograd
    d) Stalingrad   e) St. Petersburg

2. Novosibirsk is a leading manufacturing center in the Soviet region called:

    a) European Russia   b) the Far East   c) the *Povolzhye*
    d) the Russian Core   e) the Eastern Frontier

3. Which of the following leaders was a native of Georgia:

    a) Lenin   b) Stalin   c) Yeltsin
    d) Peter the Great   e) Gorbachev

4. The Shi'ite Azeri ethnic group is located in:

    a) the Southern Urals   b) the Russian Far East   c) the St. Petersburg area
    d) Azerbaijan   e) Chechnya

5. The main river of the West Siberian Plain is the:

    a) Amur   b) Volga   c) Ob
    d) Lena   e) Ural

True-False

1. Winters are long, dark, and bitter in most of Russia, but extended, warm summers make for long growing seasons. F

2. Russia is the world's largest state in population size. F

3. In 1992, most of Russia's 90 internal republics would not sign the Russian Federation Treaty, committing to cooperation in a new federal system. F

4. Abkhazia is one of Georgia's internal republics. T

5. Armenia is landlocked without a coast on either the Black or Caspian Sea. T

6. Sakhalin Island is well endowed with oil and lies off Russia's Caspian Sea coast. T

Fill-Ins

1. The _Volga_ River is most closely associated with the region known as the *Povolzhye*.

2. The U.S. state named _Alaska_ was originally purchased from Russia in 1867.

3. Georgia's political geography is plagued by _Centrifugal_ forces.

4. The city formerly called Petrograd and Leningrad is today again called _St. Petersburg_

5. The leading industrial region of the Eastern Frontier is called the _Kuzbas_.

Matching Question on Soviet Cities

D 1. *Povolzhye*
F 2. Naval base on Barents Sea
A 3. Formerly Leningrad
G 4. Formerly a leading Pacific port in Russo-Japanese trade
H 5. Capital of Sakha (the Yakut) Republic
J 6. Far East steel center
B 7. Lake Baykal
C 8. Kuznetsk Basin
E 9. Soviet Detroit
I 10. Armenian capital

A. St. Petersburg
B. Irkutsk
C. Novosibirsk
D. Volgograd
E. Nizhniy Novgorod
F. Murmansk
G. Nakhodka
H. Yakutsk
I. Yerevan
J. Khabarovsk

Essay Questions

1. It is claimed that the Russian Federation, to a very large extent, is a legacy of St. Petersburg and European Russia—not the product of Moscow and the Communist Revolution. In this context, discuss the course of Russian history and territorial evolution in the two centuries prior to 1917, and conclude your overview with a statement as to whether you concur or not with the above interpretation.

2. The Soviet experiment with a centrally-planned economy was one of the U.S.S.R.'s leading organizational enterprises. Discuss the successes that central planning achieved, its failures, and speculate where the country (assuming it had survived) might today be economically and politically had it chosen a different control structure.

3. It was often said in the now-ended postwar era that communism was an effective system—for the former U.S.S.R. Discuss the overall gains that this kind of economic system brought to the average citizen, its shortcomings, and how the country would change when (and if) a fully capitalist system is put into place.

4. More than most other national capitals and primate cities, Moscow was also the critical focal point for a huge, superpower state. Discuss the historical, cultural, physical, and economic spatial advantages that Moscow possesses that allowed it to take such control of the U.S.S.R. between the early 1920s and the end of 1991. In addition, present a synopsis of the Central Industrial Region that Moscow still anchors, reviewing the productive activities that have concentrated there and further enhanced the geographic position of the capital.

5. Compare and contrast the geographical features of Georgia, Armenia, and Azerbaijan, and evaluate the potential for future development and modernization in each country.

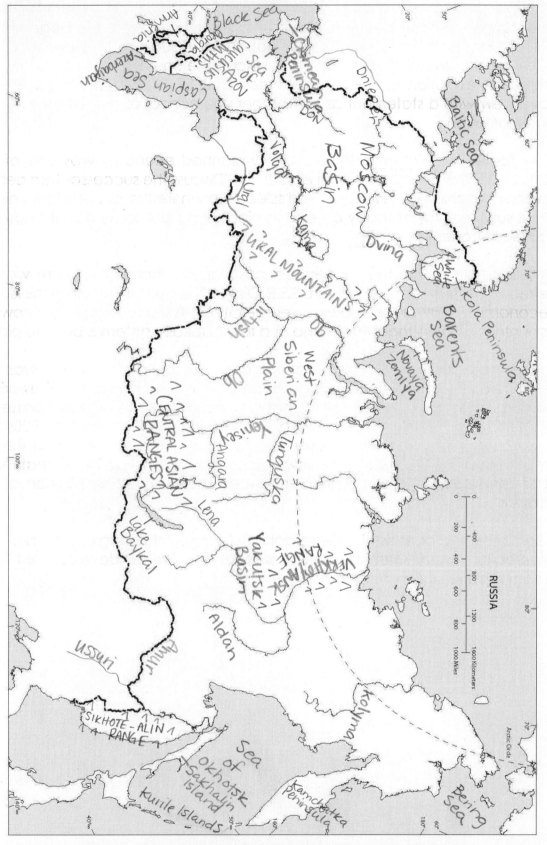

RUSSIA

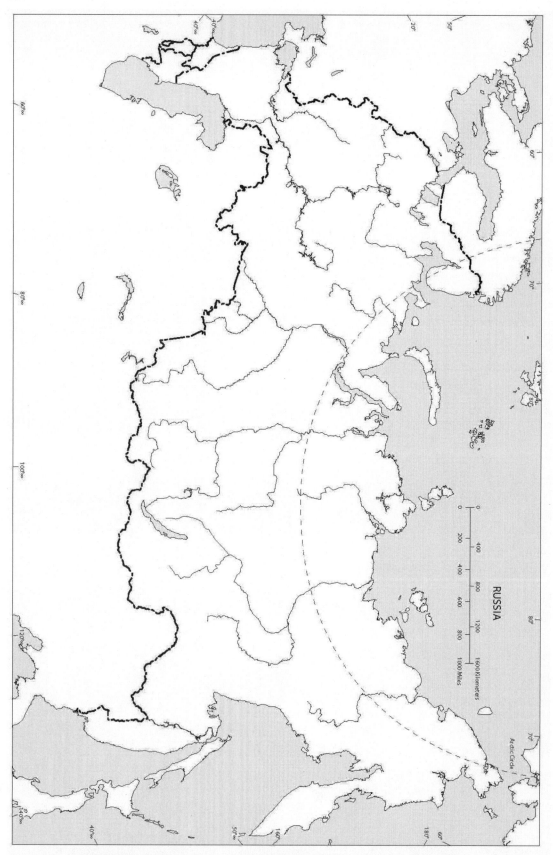

RUSSIA

Arctic Circle

0
200
400
600
800
1000 Miles

0
400
800
1200
1600 Kilometers

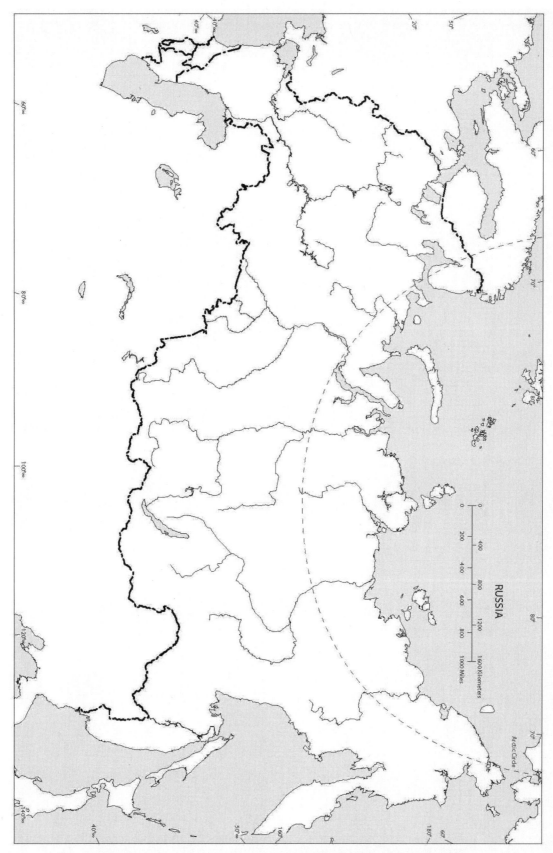

RUSSIA

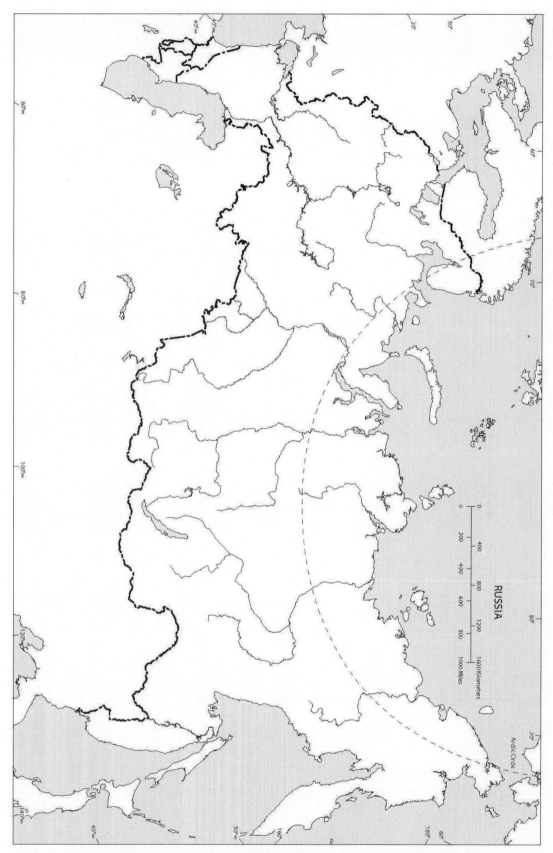

RUSSIA

CHAPTERS 3 A & B
NORTH AMERICA

OBJECTIVES OF THESE CHAPTERS

Chapters 3 A and B cover the part of the world that most of you know best, but rapid change persists in the United States and Canada as the new century unfolds. As North America's societies and economies complete their postindustrial transformation, a new geographic reality is emerging. Although this transition is not yet complete, many of its broad features have become evident since 1980 and they form the foundation for the realm's regional human geography. The dominance of urban realms and the outer suburban city on the metropolitan scene, the economic-geographic significance of the Silicon Valleys and other lineaments of the information economy, the continuing demographic shifts reported by the latest census—these are just some of the developments that are reshaping North America.

Having learned the regional geography of the United States and Canada, you should be able to:

1. Understand the similarities and differences of the U.S. and Canadian populations.

2. Grasp the essentials of the physical geography of this realm.

3. Trace the historical geography of settlement across the realm since 1800.

4. Understand the ongoing processes and patterns of metropolitan growth.

5. Grasp the emerging patterns as postindustrialism transforms culture and the economy, and urban and regional spatial change intensify.

6. Understand contemporary American economic geography—the shifting resource base, energy patterns, the dilemmas of declining "smokestack" industry, and the information economy's far-reaching spatial impacts.

7. Understand the forces that shaped contemporary Canada.

8. Understand the changing regional infrastructure of North America.

9. Locate the major physical, cultural, and economic-spatial features of the realm on an outline map.

Cultural pluralism (148)

A society composed of multiple social groups in a single state that do not mix. In Canada, cultural divisions run along ethnic and linguistic lines; in the U.S., the major social cleavage has occurred along racial and economic lines, which still fosters widespread residential segregation.

Physiographic region (148)

The clear and well-defined divisioning of a continental-scale land surface into physically uniform regions called physiographic provinces; each is marked by a general homogeneity in relief, climate, vegetation, soils, and other environmental variables.

Rain shadow effect (150)

The relative dryness in areas downwind of, or beyond, mountain ranges caused by *orographic precipitation*, wherein moist air masses are forced to deposit most of their water content.

Arid and Humid America (151)

The broad divisioning of natural environments in the United States. The average position of this north-south line is approximately 100°W longitude, located in the central Great Plains; to the east is Humid America and to the west lies Arid America (except for the narrow moist zone between the Pacific coast and its nearby parallel ranges). Humid America is generally associated with natural forest vegetation and acidic soils, whereas Arid America contains steppe or short-grass prairie vegetation and alkalinic soils.

Fossil fuels (158)

Energy resources formed by the geologic compression and transformation of ancient plant and animal organisms—*coal, petroleum (oil)*, and *natural gas*—which still account for an overwhelmingly large proportion of U.S. energy consumption.

Urban System (159)

A **hierarchical** network or grouping of urban areas within a finite geographic area, such as a country.

Productive Activities (159-160)

The major components of the spatial economy. Individual components include primary, secondary, tertiary, and quaternary economic activities.

Primary economic activity (160)

Activities engaged in the direct extraction of natural resources from the environment— particularly mining and *agriculture*.

Secondary economic activity (160)

Activities that process raw materials and transform them into finished industrial products; the *manufacturing* sector.

Tertiary economic activity (160)

Activities that engage in *services*—such as transportation, banking, retailing, finance, education, and routine office-based jobs.

Quaternary economic activity (163)

Activities engaged in the collection, processing, and manipulation of *information*.

Deindustrialization (162-163)

The loss of manufacturing due to automation and reduction of production to countries with lower wages.

Information economy (163)

Emerging economy in the United States, Canada, and a handful of other highly advanced countries as industry gives way to a high-technology productive complex dominated by services, information-related, and managerial activities.

Sunbelt (167)

The popular name given to the southern tier of the United States, which is anchored by the mega-States of California, Texas, and Florida. Its warmer climate, superior recreational opportunities, and other amenities have been attracting large numbers of relocating people and activities since the 1960s; broader definitions of the Sunbelt also include much of the western U.S., particularly Colorado and the coastal Pacific Northwest.

Gentrification (164)

Restoration of old CBD neighborhoods, often involving new construction that displaces low income local residents.

Melting Pot (167)

Traditional characterization of American society as a blend of numerous immigrant ethnic groups that over time were assimilated into a single societal mainstream. This notion always had its challengers among social scientists, and is now increasingly difficult to sustain given the increasing complexity and sheer scale of the U.S. ethnic mosaic in the twenty-first century.

Sunbelt (167)

The popular name given to the southern tier of the United States, which is anchored by the mega-States of California, Texas, and Florida. Its warmer climate, superior recreational opportunities, and other amenities have been attracting large numbers of relocating people and activities since the 1960s; broader definitions of the Sunbelt also include much of the western United States, particularly Colorado and the coastal Pacific Northwest.

Mosaic culture (167)

The ongoing fragmentation of U.S. social groups into smaller and more specialized communities, stratified not only by race and income but also by age, occupational status, and especially lifestyle.

Technopole (163; 183)

A planned techno-industrial complex (such as California's Silicon Valley) that innovates, promotes, and manufactures the products of the postindustrial informational economy.

Cross-border linkages (175)

The ties between two closely-connected localities or regions that face each other across an international boundary. These relationships are often longstanding, and intensify further as **supranationalism** proceeds (especially among the EU countries of western Europe). In North America, neighboring southwestern Ontario and southeastern Michigan are an exam-ple of such ties, which propel movements of people and goods across that stretch of the U.S.-Canada border.

Secession (175)

The act of withdrawing from a political entity, usually a **state,** as when the U.S. South tried unsuccessfully to secede from the United States in 1861 and sparked the Civil War. Successful secessions in more recent times include Singapore from Malaysia in 1965 and Slovakia from former Czechoslovakia in 1993. As **devolution** has become a force to be reckoned with around the world since the late twentieth century, the threat of secession can sometimes be a bargaining tool to gain increased local autonomy; examples include Canada (Quebec), Spain (Catalonia), and the United Kingdom (Scotland).

Megalopolis (179)

When spelled with a lower-case *m*, a synonym for **conurbation**, one of the large coalescing supercities forming in diverse parts of the world. When capitalized, refers specifically to the multimetropolitan (*Bosnywash*) corridor that extends along the northeastern U.S. seaboard from north of Boston to south of Washington, D.C.

American Manufacturing Belt (159;179)

North America's near-rectangular Core Region, whose corners are Boston, Milwaukee, St. Louis, and Baltimore. This region dominated the industrial geography of the United States and Canada during the industrial age; still a formidable economic powerhouse that remains the realm's geographic heart.

Racial Profiling (184)

The use, by police and other security personnel, of an individual's race or ethnicity in the decision to engage in law enforcement.

Pacific Rim (184)

Regional term including Japan, coastal China, South Korea, Taiwan, Thailand, Malaysia, Singapore, the United States, Canada, Australia, and Chile. This functional region is characterized by dramatic development during the past quarter-century, trade linkages, capital flow, raw material movements, urbanization, industrialization, and labor migration.

Cover the right side of the page with a sheet of paper. Uncover each line after you have attempted to answer the question in the left column. If necessary, refer to the textbook page(s) listed at the right.

Question	Answer	Page

Two Highly Advanced Societies

What are some major geographic differences between the U.S. and Canada?	The U.S. has a much bigger population, but a smaller area; the U.S. population is widely dispersed, whereas the Canadian people are highly concentrated along their southern border.	146-147

Physical Geography

What is a physiographic region?	A physically-uniform region in terms of topography, climate, vegetation, soils, and other environmental variables.	148
What is the *rain shadow effect*?	The relative dryness in areas downwind of mountain ranges, caused by *orographic* precipitation, wherein moist air masses are forced to deposit most of their water content in the highlands.	150
What are Arid and Humid America?	The dry and moist halves of the conterminous U.S., roughly divided by the transition zone along 100° W longitude.	151
What are the broadest vegetation divisions in the U.S. environment?	Arid America contains grassland vegetation; Humid America contains forest vegetation.	151
Name the five Great Lakes and their outlet to the sea.	Lakes Superior, Michigan, Huron, Erie, Ontario; the St. Lawrence River. See Fig. 3A-3, p. 150.	150

Population and Urbanization

When and how did the Native Americans most likely arrive in North America?	Over 14,000 years ago, by way of Alaska and most likely the Pacific.	152
How did the arrival of Europeans in North America in the 18th century affect Native Americans?	The Europeans ruthlessly drove the Native Americans westward from the Atlantic and Gulf coasts. After 50 years of war, what remained of Native American populations retained only 4% of US territory in the form of reservations.	152-153
When did the Industrial Revolution occur in the U.S.?	It began in the 1870s, and within 50 years America was the leading industrial power in the world.	159
What is the *core area* of the realm, and what are the four corners of the rectangular boundary that encloses this region?	The American Manufacturing Belt, cornered by the cities of St. Louis, Milwaukee, Boston, and Baltimore.	159-160
What is an outer city?	A former residential suburb that evolved into a complete and self-sufficient city, with its own businesses, industries, sports, entertainment, and other amenities.	162
Describe the geography of the information economy.	Activity concentrates in CBDs (eg NY), and suburban areas with high densities of highly skilled workers (eg Silicon Valley or Research Triangle).	163

Cultural Geography

What is the emerging *mosaic culture*?	The ongoing fragmentation of American society into a plethora of narrowly-defined communities, not only along income, racial, and ethnic lines, but also according to age and lifestyle.	167

Economic Geography

Identify: (1) primary, (2) secondary, (3) tertiary, and(4) quaternary economic activity.	(1) The extractive sector, especially mining and agriculture; (2) manufacturing; (3) the services sector; (4) the information sector.	160; 163
Name the three *fossil fuels* and their three leading North American source areas.	*Coal*—Appalachia, the northern Great Plains, Midcontinent; *petroleum (oil)*—Gulf Coast, Midcontinent, and Alaska; *natural gas*—Gulf Coast, Midcontinent, and Appalachia. See 3A-8, p. 158.	158
What are the growth industries in the information economy of the U.S.?	High-technology, white-collar, office-based activities.	163
What is a *technopole*?	A planned techno-industrial complex (like California's Silicon Valley) that innovates, promotes, and manufactures the products of the postindustrial informational economy.	163; 183

North American Regions

What is the Continental Core?	The Manufacturing Belt, now facing problems in keeping abreast of the realm-wide transition from industrial to postindustrial age.	178-179
What problems do northern New England and Atlantic Canada share?	Both are generally rural, possess difficult environments, and were historically bypassed in favor of more dynamic and fertile inland areas; although agriculture and fishing are no longer growth industries, tourism presents an opportunity in this scenic region.	179
Where is the heart of Francophone Canada?	Quebec.	180

What are the South's persistent economic problems?	Uneven development has favored certain areas, left others untouched by recent progress; much growth has occurred at the edges of the region; and Southern central cities increasingly face the problems of their Northern counterparts while their suburbs thrive.	180-181
What are the tricultural influences evident in the Southwest?	The growing Anglo influence, the persistently Hispanic flavor of local cultures, and the sporadic Native American presence.	183-184
Where is the hottest growth area of the Western Frontier?	The fastest-growing metropolis in the U.S. is the Las Vegas Valley, with its boom triggered by Las Vegas' recreation industry, and economic development fueled by the influx of high-tech and professional service firms.	185-186
What does the term Pacific Hinge refer to?	The west-cost region that forms an interface between North America and the Pacific Rim; it is the North American gateway to opportunities blossoming on the distant shores of the Pacific Basin.	184

MAP EXERCISES

Map Comparison

1. The Canadian population has always adhered closely to the country's southern border. By comparing the maps of Canada (Fig. 3B-2, p. 174), population distribution (Fig. 3A-2, p. 149). Record your observations in detail about this spatial pattern over the past two centuries.

2. Study the map of North American Manufacturing (Fig. 3A-9, p. 159), and make observations about the concentration of industrial activity in the American Manufacturing Belt with respect to other parts of the United States.

3. Study the map of North America's leading deposits of fossil fuels (Fig. 3A-8, p. 158). Learn the distributions of each energy resource by listing the major regions where coal, oil, and natural gas, respectively, are produced.

Map Construction (Use outline maps at the end of this chapter)

1. In order to become more familiar with the North American environment, place the following physical-geographic information on the first outline map:

    a. *Rivers:* Connecticut, Hudson, Delaware, Susquehanna, Potomac, Ohio, Tennessee, Mississippi, Missouri, Arkansas, Platte, Red River of the North, Rio Grande, Colorado, Snake, Columbia, Willamette, Sacramento, San Joaquin, St. Lawrence, Fraser, Okanagan

    b. *Water bodies:* Chesapeake Bay, Gulf of Mexico, San Francisco Bay, Puget Sound, Great Salt Lake, Lake Superior, Lake Michigan, Lake Huron, Lake Erie, Lake Ontario, Bay of Fundy, Gulf of St. Lawrence, Lake Winnipeg, Juan de Fuca Strait

    c. *Land bodies:* Cape Cod, Long Island, Delmarva Peninsula, Cape Hatteras, Florida Keys, Mississippi Delta, Mojave Desert, Olympic Peninsula, Vancouver Island, Grand Canyon, Alaskan Peninsula, Aleutian Islands, Newfoundland

    d. *Mountains:* Laurentians, Appalachians, Green Mountains, White Mountains, Adirondacks, Great Smoky Mountains, Ozark Plateau, Rocky Mountains, Wasatch Mountains, Sierra Nevada, Cascades, Klamath Mountains, Olympic Mountains, Pacific Coast Ranges, Black Hills

2. On the second map, political-cultural information should be entered as follows:

a. ~~Write in the name of each U.S. State and Canadian province and territory.~~ *Capital cities of each*

b. Draw in the major concentrations of minority populations, using Fig. 3B-4 (p. 176) as a guide.

3. On the third outline map, urban-economic information should be entered as follows:

a. Cities: (locate and label with the symbol ●): Boston, Hartford, New York, Buffalo, Philadelphia, Pittsburgh, Baltimore, Washington, D.C., Richmond, Norfolk, Raleigh, Charlotte, Atlanta, Charleston (S.C.), Savannah, Jacksonville, Tampa, Orlando, Miami, Mobile, Birmingham, New Orleans, Memphis, Nashville, Louisville, Cincinnati, Columbus, Cleveland, Indianapolis, Detroit, Milwaukee, Chicago, Minneapolis-St. Paul, St. Louis, Des Moines, Kansas City, Omaha, Oklahoma City, Tulsa, Houston, Dallas-Ft. Worth, San Antonio, Austin, El Paso, Albuquerque, Denver, Salt Lake City, Tucson, Phoenix, Las Vegas, San Diego, Los Angeles, San Francisco, Portland (Ore.), Seattle, Vancouver, Calgary, Regina, Winnipeg, Windsor, Toronto, Ottawa, Montreal, Quebec City, Halifax

b. *Economic regions* (identify with circled letter):

A - Silicon Valley
B - Main Street conurbation
C - Pacific Hinge
D - Alaskan North Slope
E - Research Triangle
F - Boundary between Arid and Humid America (draw in)

PRACTICE EXAMINATION

Short-Answer Questions

Multiple-Choice

1. Which of the following is not a "corner" city of the American Manufacturing Belt (Continental Core Region)?

    a) St. Louis   b) Boston   c) Milwaukee
    d) Philadelphia   e) Baltimore

2. Which city is located closest to the Canadian capital of Ottawa?

    a) Toronto   b) Windsor   c) Vancouver
    d) Detroit   e) Washington, D.C.

3. Which of the following is a secondary economic activity?

    a) iron mining   b) beer brewing   c) retail sales
    d) managing a corporation e) cotton farming

4. In which region is Las Vegas located?

    a) Pacific Hinge   b) Western Frontier   c) Southwest
    d) Continental Core   e) Northern Frontier

5. Which of the following is Canada's leading Pacific-coast city?

    a) Vancouver   b) Winnipeg   c) Alberta
    d) Ft. McMurray   e) Seattle

6. Which State lies entirely outside of the North American Core Region?

    a) New York   b) New Jersey   c) Maine
    d) Michigan   e) Maryland

True-False

1. Manitoba is one of Canada's Prairie Provinces. T

2. California's Silicon Valley is a good example of a *technopole*. T

3. Canada's population has surpassed 100 million. F

4. Florida belongs to the South region of North America. T

5. California lies entirely within Arid America. F

6. Each of the Great Lakes is bordered by Canada. F

Fill-Ins

1. The high-tech cluster that surrounds the U.S. city of _Boston_ is aligned along the circumferential Route 128 expressway.

2. The northward extension of the Sierra Nevada through the states of Oregon and Washington is called the _Cascade_ Mountains.

3. The outlet to the sea for the Great Lakes is the _St. Lawrence_ River.

4. Alaska lies wholly within the North American region called the _Northern Frontier_

5. Denver and Salt Lake City lie on opposite sides of the physiographic province known as the _Rocky Mountains_

6. The subnational political entity within Canada, a huge territory created in 1999, that covers much of the country's northeast, is called _Nunavut_.

Matching Question on States and Provinces

G	1.	Atlantic province	A.	Nunavut
I	2.	Athabasca Tar Sands	B.	Texas
K	3.	Boston	C.	California
M	4.	Grand Canyon	D.	New York
O	5.	North Slope oilfields	E.	British Columbia
B	6.	High-tech complex triangle	F.	Virginia
D	7.	Long Island	G.	Nova Scotia
C	8.	San Andreas fault	H.	Quebec
H	9.	Francophone Canada	I.	Alberta
A	10.	Newest Canadian Territory	J.	Ottawa
J	11.	Canadian capital	K.	Massachusetts
L	12.	Corn Belt	L.	Contiental Interior
E	13.	Vancouver	M.	Arizona
F	14.	Capital Beltway	N.	Louisiana
N	15.	Mississippi Delta	O.	Alaska

93

Essay Questions

1. It is often said that one of North America's greatest achievements was to overcome the "tyranny" of distance. Discuss how the United States and Canada overcame their physical barriers to allow their transcontinental economies and societies to emerge. Compare and contrast this experience in effective long-distance spatial organization with that of Russia since 1900 (referring back to the preceding chapter if necessary).

2. Imagine that you are the chief executive of a successful young company that manufactures software for the newest high-speed computers. Where would you locate your company and why? Establish a "short list" of three possible sites, and choose one by a process of elimination in which your decision is built upon the most important locational variables for your plant and its highly-skilled work force.

3. Most of the western half of the United States is part of Arid America. Explain why this dry environment exists, and what its consequences are for climate, vegetation, and soils. Also, discuss the implications of this situation for the human use of the Earth, and speculate about what kind of economy might have arisen had the country been settled from the *Pacific coast* moving inland to the east.

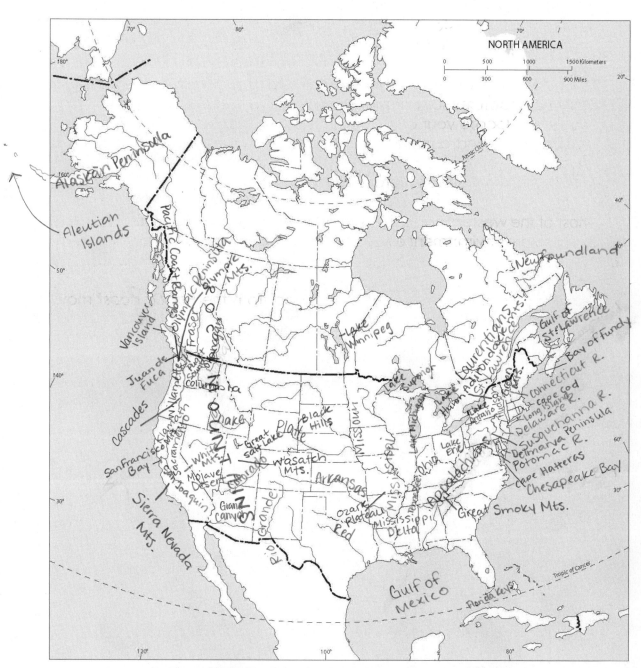

NORTH AMERICA

500   1000   1500 Kilometers
300   600   900 Miles

Aleutian Islands

Alaskan Peninsula

Pacific Coast Ranges

Vancouver Island

Olympic Peninsula

Olympic Mts.

Fraser

Puget Sound

Juan de Fuca

Cascades

Columbia

Willamette

Sacramento

San Francisco Bay

San Joaquin

Sierra Nevada Mts.

Mojave Desert

White Mts.

Grand Canyon

Colorado

Snake

Great Salt Lake

Wasatch Mts.

Platte

Black Hills

Rio Grande

Red

Ozark Plateau

Arkansas

Missouri

Mississippi

Mississippi Delta

Ohio

Tennessee

Lake Michigan

Lake Erie

Lake Superior

Lake Huron

Lake Winnipeg

Lake Ontario

St. Lawrence R.

Laurentians

Adirondacks

Newfoundland

Gulf of St. Lawrence

Bay of Fundy

Connecticut R.

Cape Cod

Long Island

Delaware R.

Susquehanna R.

Delmarva Peninsula

Potomac R.

Cape Hatteras

Chesapeake Bay

Appalachians

Great Smoky Mts.

Gulf of Mexico

Florida Keys

Tropic of Cancer

Arctic Circle

US/CANADA

U.S. Capitals

1. Olympia
2. Oklahoma City
3. Pierre
4. Bismarck
5. St. Paul
6. Des Moines
7. Jefferson City
8. Little Rock
9. Madison
10. Springfield
11. Lansing
12. Jackson
13. Nashville
14. Frankfort
15. Indianapolis
16. Columbus
17. Montgomery
18. Atlanta
19. Richmond
20. Charleston
21. Harrisburg
22. Albany
23. Augusta
24. Conco
25. Montpelier
26. Boston
27. Providence
28. Hartford
29. Trenton
30. Dover
31. Annapolis

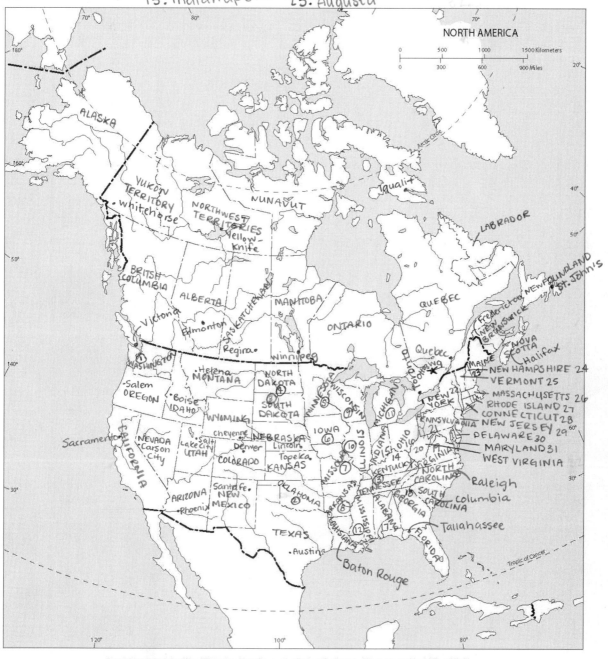

NORTH AMERICA

Copyright © 2010 John Wiley & Sons, Inc. From Geography: Realms, Regions, and Concepts, 14e by deBlij and Muller

97

NORTH AMERICA

NORTH AMERICA

CHAPTERS 4 A&B
MIDDLE AMERICA

OBJECTIVES OF THESE CHAPTERS

Chapters 4 A and B are a survey of Middle America (Mexico, Central America, and the Greater and Lesser Antilles that constitute the islands of the Caribbean Basin). Following an introduction, the sequential influences of the Amerindian (Mesoamerican) civilizations and the Hispanic colonizers are reviewed, and their various cultural collisions are evaluated in the shaping of contemporary society. The regional structure of the realm is initially presented within the useful "Mainland-Rimland" framework (set against the complex European colonial legacy). Each major region is then treated, and the Caribbean—underscoring cultural and economic geography and the impact of tourism, Mexico—highlighting social and economic geography as well as plans for continuing development; Central America—profiling each of its 7 republics, and emphasizing spatial dimensions of the instabilities that continue to plague this region.

Having learned the regional geography of Middle America, you should be able to:

1. Understand which components make up the Middle American realm and their differing physiographies.

2. Describe the major contributions of the Maya, Aztecs, and the Spanish in shaping the contemporary cultural and social geography of the realm (and the African and other European colonial infusions in much of the Rimland portion of Middle America).

3. Differentiate between Mainland and Rimland Middle America in terms of political, cultural, and economic regional geography.

4. Understand the geography of Mexico, especially its development opportunities as well as its problems related to economic inequalities and population growth.

5. Understand the altitudinal zonation of environments that mark the economic and settlement geographies of Middle and South America.

6. Understand the geographic patterns of Central America, including the challenges facing each of its republics.

7. Understand the geographic patterns of the Caribbean Basin, its evolution, and its prospects for future change.

8. Locate the major physical, cultural, and economic-spatial features of the realm on an outline map.

Chapter 4A

NAFTA (194)

The **free-trade area** launched in 1994 involving the United States, Canada, and Mexico.

Isthmus (197)

Performs the same function as a land bridge, but is usually shorter in length and narrower in width. On the mainland Middle American land bridge, the South American end is the isthmus of Panama.

Land Bridge (197)

A narrow isthmian link that connects two larger landmasses, such as the 3800-mile-long (6000-km-long) mainland of Middle America that connects North and South America between the U.S. border and the southeastern end of the Panamanian isthmus.

Archipelago (197)

A set of islands grouped closely together, usually elongated into a chain.

Greater Antilles (197)

The larger islands of the northern Caribbean that encompass Cuba, Hispaniola (containing Haiti and the Dominican Republic), Puerto Rico, and Jamaica.

Lesser Antilles (197)

The smaller-island arc of the eastern Caribbean, stretching southward from the Virgin Islands to Trinidad near the South American coast. Region can also be extended northwestward toward Florida to include the Bahamas island chain.

Maquiladora (197; 214-map)

The term given to modern industrial plants in Mexico's northern (U.S.) border zone. These foreign-owned factories assemble imported components and/or raw materials, and then export finished manufactures, mainly to the United States.

Hurricane Alley (198)

The most frequent pathway followed by tropical storms and **hurricanes** over the past 150 years in their generally westward movement across the Caribbean Basin. Historically, hurricane tracks have bundled most tightly in the center of this route, most often affecting the Lesser Antilles between Antigua and the Virgin Islands, Puerto Rico, Hispaniola (Haiti/Dominican Republic), Jamaica, Cuba, southernmost Florida, Mexico's Yucatán, and the Gulf of Mexico.

Altitudinal Zonation (199)

Vertical regions defined by physical-environmental zones at various elevations, particularly in the highlands of South and Middle America. See Fig 4A-4.

Tierra caliente (199)

The lowest of five vertical zones into which the settlement of highland Middle and South America is divided according to elevation. The *caliente* is the hot humid coastal plain and adjacent slopes up to 2500 feet (750 meters) above sea level. The natural vegetation is the dense and luxuriant tropical rainforest; the crops are tropical, including bananas. See Fig. 4A-4, p. 198.

Tierra templada (199)

The second altitudinal zone in highland Middle and South America, between 2500 and 6000 feet (750 and 1850 meters). This is the "temperate" zone, with moderate temperatures compared to the *tierra caliente*. Crops include tobacco, coffee, corn, and some wheat. See Fig. 4A-4, p. 198.

Tierra fría (199)

The third altitudinal zone in highland Middle and South America, from about 6000 feet (1850 meters) up to the tree line at nearly 12,000 feet (3600 meters). Coniferous trees stand here; upward they change into scrub and grassland. There are also important pastures within the *fría*, and wheat, potatoes, and barley can be cultivated. See Fig. 4A-4, p. 198.

Tierra helada (149)

The fourth settlement zone in highland South America, extending upward from about 12,000 to 15,000 feet (3600-4500 meters). This altitudinal zone lies above the tree line, and is so cold and barren that it can only support the grazing of hardy livestock and sheep. See Fig. 4A-3, p. 149.

Tierra nevada (199)

This uppermost zone is above the snow line, which lies at approximately 15,000 feet (4500 meters), and is referred to as the "frozen land." See Fig. 4A-4, p. 198.

Tropical Deforestation (199)

The clearing and destruction of tropical rainforests to make way for expanding settlement frontiers and the exploitation of new economic opportunities.

Culture hearth (199)

A source area or innovation center from which cultural traditions are transmitted.

Mesoamerica (199)

Anthropological label for the Middle American culture hearth.

Plaza (201)

Central market square that was the focus of Spanish New World towns, containing church and government buildings.

Mainland-Rimland framework (203)

Augelli's framework that recognizes a Euro-Amerindian Mainland and a Euro-African Rimland in Middle America.

Hacienda (204)

Literally, a large estate in a Spanish-speaking country. Sometimes equated with plantation, but there are important differences between these two types of agricultural enterprise and rural land tenure. Many have now been divided into smaller holdings or reorganized as cooperatives.

Plantation (203)

A large estate owned by an individual, family, or corporation and organized to produce a cash crop. Almost all plantations were established within the tropics.

Chapter 4B

Acculturation (213)

Cultural modification resulting from intercultural borrowing. In cultural geography, the term refers to the change that occurs in the culture of indigenous peoples when contact is made with a society that is technologically superior.

Small-island developing economies (205)

The additional disadvantages faced by lower-income island-states because of their often small territorial size and populations as well as overland **inaccessibility**. Limited resources require expensive importing of many goods and services; the cost of government operations per capita are higher; and local production is unable to benefit from **economies of scale**. The eastern Caribbean islands of the Lesser Antilles are a good example.

Acculturation (213-214)

Cultural modification resulting from intercultural borrowing. In cultural geography, the term refers to the change that occurs in the culture of indigenous peoples when contact is made with a society that is technologically superior.

Transculturation (214)

Two-way cultural borrowing that occurs when different cultures of approximately equal complexity and technological level come into close contact. In *acculturation*, by contrast, an indigenous society's culture is modified by contact with a technologically more advanced society.

Peones (214)

Landless, indebted serfs, who lived in Mexico in the 19th century.

Ejido (214)

Communally-owned cooperative farmlands in central and southern Mexico; former *hacienda* lands.

Mestizo (Ladino) (220)

A person of mixed white and Amerindian ancestry.

Mulatto (224)

A person of mixed white and African ancestry.

Social Stratification (224)

In a layered or stratified society, the population is divided into a **hierarchy** of social classes. In an industrialized society, the working class is at the lower end; **elites** that possess capital and control the means of production are at the upper level. In the traditional **caste system** of Hindu India, the "untouchables" form the lowest class or caste, whereas the still-wealthy remnants of the princely class are at the top.

Cover the right side of the page with a sheet of paper. Uncover each line after you have attempted to answer the question in the left column. If necessary, refer to textbook page(s) listed at the right.

Question	Answer	Page
Middle American Characteristics		
What is the spatial extent of this realm?	Mexico and Central America—from the U.S. border to the northern edge of South America—plus all of the Caribbean islands to the east.	192-193
What is the difference between Central and Middle America?	*Central America* is the mainland between Mexico and South America, containing the 7 republics of Belize, Guatemala, Honduras, El Salvador, Nicaragua, Costa Rica, and Panama. Besides all of Central America, *Middle America* also includes Mexico and the Caribbean Basin.	193-194
Why is Middle America more culturally diverse than South America?	African and Asian ancestries prevail beside those of European background; the Amerindian cultural contribution is greater; the Caribbean is a region of especially complex cultural pluralism.	193-194
What is a *land bridge*?	A narrow isthmian link between two large landmasses.	197
Mesoamerican Legacy		
Name the five altitudinal zones of human settlement in highland Middle and South America.	*Tierra caliente* (sea level to 2500 feet), *tierra templada* (2500-6000 feet), *tierra fria* (6000-12,000 feet), *tierra helada* (12,000-15,000 feet), and the *tierra nevada* (15,000 feet and above). See Fig. 4A-4, p. 198.	199
Environmentally, where is most of the region's population concentrated?	In the *templada* zone of the highlands, toward the Pacific side of the land bridge.	199

What is *Mesoamerica*?	The Middle American culture hearth that stretched from northern Mexico southeast to central Nicaragua.	199
What is unique about where the Maya culture hearth developed?	It is the only one that arose in the lowland tropics.	200
Describe some of the Maya's accomplishments.	Urbanization, pyramids, spectacular palaces, stone carvings and other artwork, mathematics, astronomy, calendrics, advanced agriculture, and a wide trading network.	200
What was the central area of the Aztec's empire?	The Aztec state centered in the Valley of Mexico, headquartered at the city of Tenochtitlán.	200
Which crops did the Amerindians of Meso-america contribute to the world?	Corn (maize), various kinds of beans, the sweet potato, the tomato, cacao, squash, and tobacco.	201
What kinds of agriculture were emphasized after the Spanish conquest?	The keeping of livestock, especially cattle and sheep—which competed with the growing of the subsistence crops of the conquered Amerindians.	201
What was the most far-reaching change in the cultural landscape brought by the Spaniards?	The resettlement of the Amerindians from rural land into villages and towns laid out and controlled by the conquerors; these towns were administrative centers for tax collection and labor recruitment, especially for mining.	201
How were Spanish New World towns organized?	The focus of the town was the central plaza, where church and government buildings were located. The surrounding streets were arranged in an easily defensible gridiron pattern. See Fig. 4A-5, p. 201.	201

Mainland and Rimland

What is the difference between the cultures of Mainland and Rimland Middle America?	The Mainland is dominated by a Euro-Amerindian cultural heritage; the Rimland is dominated by a Euro-African cultural heritage.	203-204

112

What are the geographic differences between Mainland & Rimland?	The Rimland was an area of sugar and banana plantations, high accessibility, seaward exposure, & maximum cultural contact & mixture; the Mainland was removed from these contacts, the region of the hacienda, more self-sufficient, & less dependent on outside markets.	203-204
What are some of the characteristics of Middle American plantations?	They are located in the tropical coastlands and islands; they produce single export crops; foreign ownership and profit outflow dominate; labor is seasonal and has often been imported; "factory in the field" methods are far more efficient than those of the hacienda.	205

Mexico

How strong is the Amerindian imprint on Mexican culture?	Extremely strong: 62 percent of the population is mestizos, 21 percent are predominantly Amerindian, about 7% are fully Amerindian, and only 10 percent are European.	213
Did the 1910 Revolution achieve its goals?	For the most part, yes: the *haciendas* were redistributed, many into communally-owned *ejidos*; the Revolution also resurrected the Amerindian cultural contribution to Mexican life.	214
Where is most of Mexico's oil production located?	Along the central and southern Gulf Coast, especially around the city of Villahermosa & the nearby Bay of Campeche. See Fig. 4B-2, p. 211.	211
What are *maquiladoras*?	Modern industrial plants in Mexico's northern (U.S.) border zone. These foreign-owned factories assemble imported components and/or raw materials, and then export finished manufactures, mainly to the United States. Most import duties are minimized, bringing jobs to Mexico and the advantages of low wage rates to the foreign entrepreneurs.	214

Central America

Name the seven republics of Central America.	Guatemala, Honduras, Belize, El Salvador, Nicaragua, Costa Rica, and Panama. See Fig. 4B-8.	219-map

What basic conflicts divide the population of Central America?	Amerindian and mestizo population clusters often clash; also, there is a huge gulf between the privileged and the poor, causing resentment and violence.	219
What is a *mestizo*?	A person of mixed white and Amerindian ancestry.	220
Which republic was struck by Hurricane Mitch in 1989? Which republic contains a coffee growing area called the Valle Central?	Nicaragua; Costa Rica.	222
What is the current status of the Panama Canal?	The U.S. withdrawal was completed at the end of 1999 and the Canal turned over to Panama. Expansion plans for the canal will boost traffic and business opportunities.	223

The Caribbean Basin

Why do the export crops and the mineral resources of the islands provide so modest an income for the region, and why does wide-spread poverty persist?	These commodities face severe competition from many other disadvantaged countries, and are not established on a scale that could improve local living standards; most islanders therefore live in poverty and eke out a subsistence existence from a small plot of poor land.	224-225
Why is there a large population of Asian peoples in the Caribbean?	During the 19th century, emancipation of slaves and ensuing labor shortages brought immigrants (mostly indentured servants) from distant locations, including China and India.	224
Which islands had sizeable influxes of Asian immigration in the 19th century?	Cuba, Trinidad, Jamaica, and Guadeloupe.	225
Which countries comprise the Netherlands Antilles?	Curaçao, Aruba, and St. Maarten.	230

MAP EXERCISES

Map Comparison

1. The map of Middle America's regions (Fig. 4B-1, p. 208-209) clearly offsets each of the geographic components of this realm. Compare and contrast the environ-ments of these regions, basing your observations on the world maps of landscapes (Fig. G-1), and climates (G-7) in the introductory chapter.

2. Compare Figs. 4A-6 and 4A-7 (pp. 202-204) and make observations about the colonial influence on the Mainland/Rimland split. Why is the Central American east coast included in the Rimland rather than the Mainland region? What generalizations can be made about the composition of present Caribbean populations based on the European countries that once ruled them?

3. A significant recent development in Mexico's manufacturing geography is the rise of the *maquiladoras*. Compare the maps of maquiladora location Fig. 4B-6 (p. 214), raw materials Fig. 4B-2 (p. 211), and the internal States of Mexico Fig. 4B-4 (p. 213). Comment on the advantages that the northern border holds for industrial activity, and on what the highly clustered location of *maquiladoras* means for the various regional economies of Mexico.

Map Construction *(Use outline maps at the end of this chapter)*

1. In order to familiarize yourself with Middle American physical geography, place the following on the first outline map:

   a. *Water bodies*: Caribbean Sea, Gulf of Mexico, Bay of Campeche, Gulf of California (Sea of Cortés), Gulf of Tehuantepec, Gulf of Honduras, Gulf of Panama, Gulf of Darien, Gulf of Fonseca, Panama Canal, Rio Grande River, Lake Nicaragua, Straits of Florida, Windward Passage

   b. *Land bodies and features*: Greater Antilles, Lesser Antilles (including the Bahamas Islands), Leeward Islands, Windward Islands, Virgin Islands, Baja California, Yucatán Peninsula, Isthmus of Panama, Hispaniola, Florida Keys, Sierra Madre Occidental, Sierra Madre Oriental, Valley of Mexico, Sonora Desert, Isthmus of Tehuantepec

2. On the second map, enter the name of every country that is shown, including all of the mainland republics and as many island-states as possible.

3. On the third outline map, urban-economic information should be entered as follows:

a. *Capital cities* (locate and label with the symbol *): Mexico City, Belmopan, Guatemala City, San Salvador, Tegucigalpa, Managua, San José, Panama City, Havana, Kingston, Nassau, Port-au-Prince, Santo Domingo, San Juan, Fort-de-France, Port of Spain, Willemstad

b. *Other cities* (locate and label with the symbol ●): Ciudad Juarez, Tijuana, Monterrey, Torreón, Chihuahua, Durango, Zacatecas, Guadalajara, Tampico, Veracruz, Acapulco, Oaxaca, Villahermosa, Mérida, Cozumel, Belize City, Quezaltenango, San Pedro Sula, San Miguel, León, Bluefields, Limon, Colón, Santiago de Cuba, Mariel, Guantanamo, Montego Bay, Puerto Plata, Mayaguez

c. *Economic Regions* (identify with circled letter):

A - Curaçao
B - Major area of *ejidos* today
C - Bay of Campeche oilfield
D - Panama Canal
E - Mexican gold placering area
F - *Maquiladora* zone

PRACTICE EXAMINATION

Short Answer Questions

Multiple-Choice

1. The racial term applied to people of mixed white and Amerindian ancestry is:

   a) *mulatto* b) *caliente* c) *ejido*
   d) *mestizo* e) *amerindiano*

2. Which mainland republic was embroiled in civil war throughout the 1980s?

   a) Nicaragua b) Panama c) Honduras
   d) Mexico e) Dominican Republic

3. The most populous of all the Middle American countries is:

   a) Nicaragua b) Panama c) Mexico
   d) Honduras e) Guatemala

4. Which Mainland country does not contain a portion of the Rimland?

   a) El Salvador b) Belize c) Honduras
   d) Nicaragua e) Panama

5. Which part of Mexico is most closely associated with the ancient Maya culture?

   a) Baja California b) Valley of Mexico c) the Yucatán Peninsula
   d) Hispaniola e) Rio Grande Valley

6. Which Central American capital is located on the Pacific?

   a) San Jose b) Mexico City c) Belmopan
   d) Guatemala City e) Panama City

True-False

1. The core area of the Aztec state was located in what is still the core area of Mexico today.

2. The large island of Trinidad is located in the Greater Antilles.

3. Costa Rica is a U.S. enemy that is a haven for communist insurgencies throughout Middle and South America.

4. The Chiapas rebellion occurred in southern Mexico.

5. The *tierra nevada* altitudinal zone does not occur in Central America.

6. Before Castro's revolution, Cuba was a commonwealth of the United States.

Fill-Ins

1. The gridiron street pattern was first introduced to Middle America by Europeans from the country of _Spain_.

2. The lowest-lying altitudinal zone of agricultural activity, extending from sea level to an elevation of 2500 feet (750 meters), is the *tierra* _caliente_.

3. The Yucatán Peninsula is a part of the country of _mexico_.

4. The largest island of the Greater Antilles is _Cuba_.

5. The country most widely devastated by Hurricane Mitch in 1998 was _Honduras_.

6. The only Middle American country with a direct land link to the South American continent is _Panama_.

Matching Question on Middle American Countries

_O_ 1.	U.S. backed government in civil war	A. Mexico
_K_ 2.	Former bastion of leftist insurgency	B. Puerto Rico
_L_ 3.	"Switzerland of Central America"	C. Dominican Republic
_N_ 4.	Large East Indian population	D. Curaçao
_C_ 5.	Eastern half of Hispaniola	E. Jamaica
_F_ 6.	Former Danish colony	F. U.S. Virgin Islands
_B_ 7.	Possibility of U.S. statehood	G. Honduras
_M_ 8.	The largest Caribbean island	H. Haiti
_E_ 9.	Bauxite (aluminum ore) mining	I. Belize
_J_ 10.	Formerly belonged to Colombia	J. Panama
_D_ 11.	Former a Dutch colony	K. Nicaragua
_H_ 12.	Western half of Hispaniola	L. Costa Rica
_G_ 13.	Hit the hardest by Hurricane Mitch	M. Cuba
_A_ 14.	Part of Maya culture hearth	N. Trinidad and Tobago
_I_ 15.	Formerly British Honduras	O. El Salvador

Essay Questions

1. One of this realm's most important regional frameworks is the Mainland/Rimland scheme. Compare and contrast the physical, economic, and cultural geographies of each of these components of Middle America, and show how its historical evolution proceeded in a direction that poses a challenge to regional integration and unity in the future.

2. Discuss the European impact on the shaping of the cultural and political geographies of the Caribbean Basin today. Why is this region trapped in a cycle of poverty that is rooted in the economic system inherited from the colonial era?

3. Describe the fivefold vertical zonation of settlement environments in highland Middle and South America, including the agricultural geography of each zone.

4. Few countries in the developing world possess the opportunities and challenges facing Mexico. Discuss those aspects of Mexican geography that offer a real potential for meaningful progress in living standards in the foreseeable future, and weigh them against the awesome problems the country confronts in the short- and longer-term future.

5. Much of Central America's turmoil has been rooted in this region's past. Trace the historical geography of the region, paying attention to Amerindian, Spanish, and other cultural influences. What new economic prospects brighten Central America's future, and what obstacles must be overcome in order for real progress to occur?

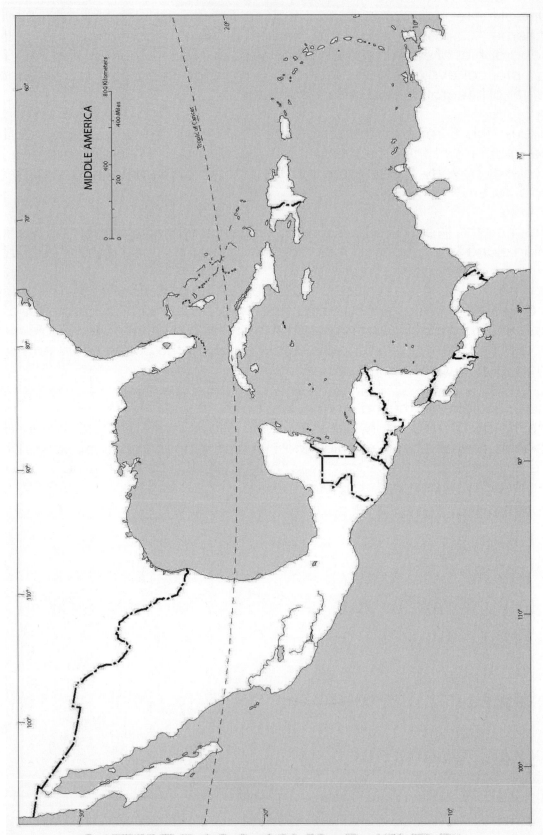

MIDDLE AMERICA

Tropic of Cancer

0   200   400   600   800 Kilometers
0   200   400 Miles

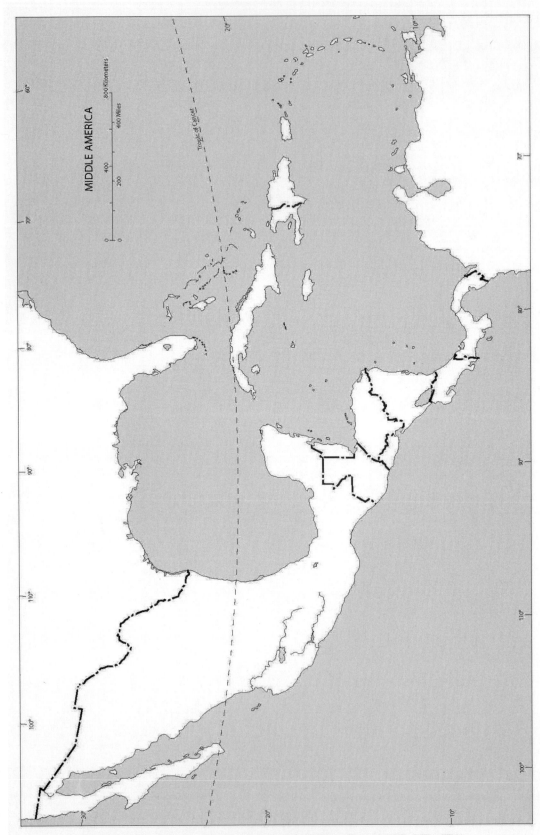

MIDDLE AMERICA

Tropic of Cancer

800 Kilometers
400 Miles

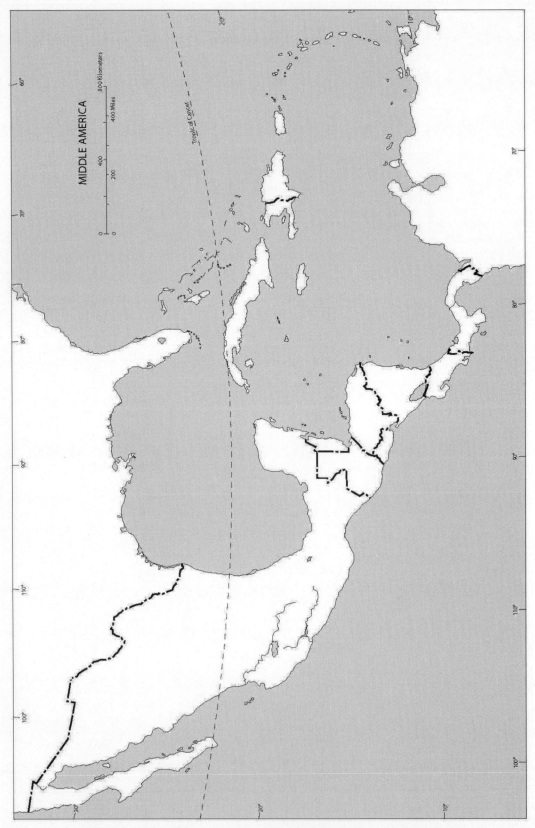

MIDDLE AMERICA

Tropic of Cancer

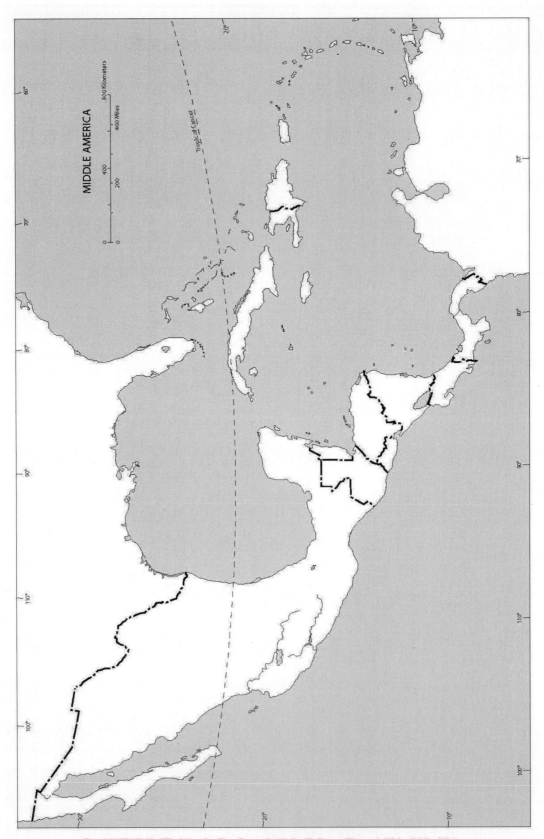

MIDDLE AMERICA

Tropic of Cancer

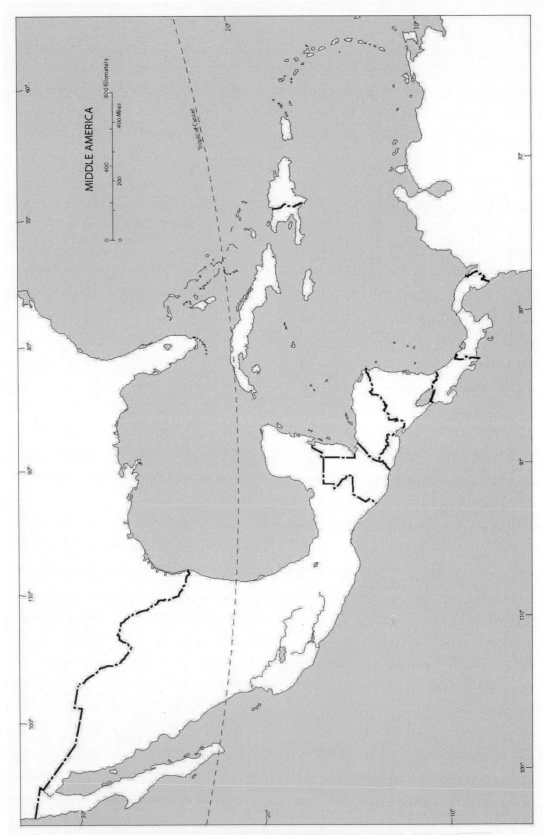

MIDDLE AMERICA

800 Kilometers
400 Miles

400
200

Tropic of Cancer

CHAPTERS 5 A&B
SOUTH AMERICA

OBJECTIVES OF THESE CHAPTERS

Chapters 5 A and B treat South America, a realm of vast potential but also of frustrations in the effort to move itself forward. After a general introduction to the chapter, South America's historical geography is traced in some detail, with emphasis on its component culture areas. Urbanization is introduced, highlighting a model of "Latin" American city structure. A review of South America's regions follows, presented within a four-part framework covering the Caribbean North, the Andean West, the mid-latitude Southern Cone, and giant Brazil.

Having learned the regional geography of South America, you should be able to:

1. Understand the major physiographic features of the realm.

2. Grasp the essentials of the historical geography of South America.

3. Understand the various influences that have shaped this realm's culture areas.

4. Deepen your understanding of Middle and South American urbanization, building on the Griffin-Ford model introduced on text pp. 248-249.

5. Describe the broad regionalization pattern of the realm, with reference to cultural patterns, historical, and political developments.

6. Understand the geographic essentials of each South American republic, including its resources, agricultural patterns, natural environments, and economic development potential.

7. Locate the leading physical, cultural, and economic-spatial features of the realm on an outline map.

GLOSSARY

Chapter 5A

Unity of place (234)

The great natural scientist Alexander von Humboldt's notion that in a particular locale or region intricate connections exist among climate, geology, biology, and human cultures. This laid the foundation for modern geography as an *integrative discipline* marked by a spat

Amerindian (235)

Person ethnically linked to the people who inhabited the Americas before Europeans.

Altiplano (235-236)

High-elevation plateau or basin between even higher mountain ranges; Andean *altiplanos* often lie at altitudes in excess of 10,000 feet (3000 m).

Land alienation (236)

The process by which Spanish invaders took over Amerindian lands in order to form large *haciendas*.

Plural society (240)

A society in which two or more population groups, each practicing its own culture, live adjacent to one another without mixing inside a single state.

Commercial agriculture (241)

For-profit agriculture.

Subsistence agriculture (241)

Minimum-life-sustaining agriculture.

Uneven Development (242)

The notion that economic development varies spatially, a central tenet of **core-periphery relationships** in realms, regions, and lesser geographic entities.

Free Trade Area of the Americas (FTAA) (244)

The ultimate goal of supranational economic integration in North, Middle, and South America: the creation of a single trading block that would involve every country in the Western Hemisphere between the Arctic and Cape Horn.

Urbanization (245)

The process of urbanization involves the movement to, and the clustering of, people in towns and cities.

Rural-to-urban migration (245)

The dominant migration flow from countryside to city that continues to transform the world's population, most notably in the less advantaged geographic relams.

Megacity (247)

City whose population exceeds 10 million.

"Latin" American City model (248-249)

The Griffin-Ford model of Middle and South American intraurban spatial structure, discussed on pp. 248-249, and diagrammed on p. 248.

Plaza (248)

The old hub and focus of the Middle and South American city, the open central square flanked by the main church and government buildings.

Informal sector (249)

Dominated by unlicensed sellers of goods and services, the primitive form of capitalism found in many developing countries that takes place beyond the control of government.

Barrio (favela) (249)

Term meaning "neighborhood" in Spanish. Usually refers to an urban community in a Middle or South American city; often slums are known as *barrios* or in Brazil as *favelas*.

Chapter 5B

Insurgent state (257)

Territorial embodiment of a successful guerrilla movement. The establishment by anti-government insurgents of a territorial base in which they exercise full control; thus a state-within-a-state.

Failed state (257)

A country whose institutions have collapsed and in which anarchy prevails.

Llanos (258)

Savanna-like grasslands, especially of the low-lying Orinoco Basin in Venezuela as well as neighboring Colombia.

Tierra caliente (see Chapter 5, p. 199)

The lowest of four vertical zones into which the settlement of highland South America is divided according to elevation. The *caliente* is the hot humid lowland and adjacent slopes up to 2500 feet (750 m) above sea level. The natural vegetation is the dense and luxuriant tropical rainforest; the crops are bananas, sugar, cacao, and rice in the lower areas and coffee, tobacco, and corn along the somewhat higher slopes.

Tierra templada (see Chapter 5, p. 199)

The second altitudinal zone in highland South America, between 2500 and 6000 feet (750 and 1850 m). This is the "temperate" zone, with moderate temperatures compared to the *tierra caliente*. Crops include tobacco, coffee, corn, and some wheat.

Tierra fría (see Chapter 5, p. 199)

The third altitudinal zone in highland South America, from about 6000 feet (1850 m) up to nearly 12,000 feet (3600 m). Coniferous trees stand here; upward they change into scrub and grassland. There are also important pastures within the *fría*, and wheat can be cultivated. Several major population clusters in the Andes lie at these elevations.

Tierra helada (see Chapter 5, p. 199)

The fourth settlement zone in highland South America, extending upward from about 12,000 to 15,000 feet (3600-4500 m). This altitudinal zone is so cold and barren that it can only support the grazing of hardy livestock and sheep.

Tierra nevada (see Chapter 5, p. 199)

Above the snow line, lying at approximately 15,000 feet (4500 m), this is the uninhabitable "frozen land."

Oriente (262)

Literally "the east," refers to the jungly lowlands of Peru and Ecuador to the east of the Andes that are sparsely populated but contain petroleum deposits.

Altiplanos (263)

High elevation basins and valleys of Peru and coastal northern Chile.

Pampa (266-267)

Literally the word means "plain." The physiographic subregion of east-central Argentina that is the country's leading crop-and-livestock-producing region.

Elongation (270)

A state whose territory is decidedly long (at least six times longer than its width), which often creates external political, internal administrative, and general economic problems. Chile is a classic example.

BRICs (273)

Acronym for the four biggest emerging national markets in the world today—**B**razil, **R**ussia, **I**ndia, and **C**hina.

Sertão (277)

The dry inland back-country of northeastern Brazil.

El Niño (277)

A periodic, large-scale, abnormal warming of the sea surface in the low latitudes of the eastern Pacific Ocean that has global implications, disturbing normal weather patterns in many parts of the world, especially South America.

Forward capital (279)

A capital city located near a sensitive zone with a neighboring state or a frontier that a country wishes to develop; such a statement concerning the push to the empty interior of Brazil was made in the 1950s, when the government decided to relocate its headquarters from Rio de Janeiro to Brasília.

Cerrado (280)

A biologically diverse area of savanna in the central-western Brazil subregion, now threatened by expansion of industrial-scale soybean production.

Growth pole (280)

An urban center with certain attributes that, if augmented by investment support, will stimulate regionwide economic development of its hinterland.

Cover the right side of the page with a sheet of paper. Uncover each line after you have attempted to answer the question in the left column. If necessary, refer to textbook page(s) listed at right.

Question	Answer	Page
South American Characteristics		
How does South America's longitudinal position differ from North America's?	It lies considerably farther to the east, closer to Africa, but facing a much wider Pacific Ocean on the west.	234
What is the realm's dominant physio-graphic feature?	The Andes Mountains, which form an imposing north-south barrier along South America's entire west coast.	234
Where are modern South America's largest population concentrations situated?	In the coastal east and the north of the realm.	235
Historical Geography		
Who were the Incas?	Descendants of ancient peoples who created a major civilization in the northern Andes around A.D. 1300, centered in Peru's Cuzco Basin.	236
List some of the Incas' major achievements.	Most of all, the political integration of Andean South America; a splendid circulation system for goods and ideas; administration of a complex social and economic system.	236
How and when did the Spaniards conquer the Incas?	In 1533, Pizarro led a small band of soldiers into Cuzco and deposed the Inca Empire.	236

How did Spain and Portugal resolve their 15th century disputes over South American territory?	The 1494 Treaty of Tordesillas, mediated by the pope, gave Spain all land west of the 50th meridian and Portugal all territory to the east.	237
Why did the Portuguese import so many Africans to coastal Brazil?	They opted for a Caribbean-style plantation economy, which required the services of millions of African slaves.	239
How did South America's 19th century independence movement spread?	Argentina and Chile, farthest from Peru, began it and Bolivár pushed down from New Granada to the north—by 1824, the Spaniards were driven out.	238-239

Cultural/Economic Geography

What makes the pattern of South American agriculture unusual?	Commercial and subsistence agriculture exist side by side to a greater degree than in any other realm (see Fig. 5A-5).	241
How well unified is the South American realm?	In the past, surprisingly little interaction occurred among the realm's countries, but there are signs of change as a new era of cooperation appears to be opening. Mutually advantageous trade is the catalyst for international cooperation.	244

Urbanization

How fast is South American urbanization increasing?	Very rapidly: the realm is 82% urban today; the urban areas have increased annually by 5% since 1950, the rural areas by less than 2%.	245
Why does the migration toward cities remain so high today?	*Push* factors are rural poverty and lack of land reform; reinforcing *pull* factors are perceived employment opportunities, education, medical care, and a more exciting pace of life.	245
Name the land-use zones of the "Latin" American city model.	The CBD, the commercial spine, the elite residential sector, the zone of maturity, the zone of *in situ* accretion, the zone of peripheral squatter settlements, & the disamenity sector containing the slums. Layout shown in Fig. 5A-8.	248-249

139

Caribbean, Northern South America

Which states make up this region? What forces bind them together?	Venezuela, Colombia, and the three Guianas. Common coastal location; the legacy of a tropical plantation culture and economy; large black and Asian minorities. See Fig. 5B-2, p. 256.	255
What is an insurgent state?	The territorial base in which a guerrilla movement exercises full control; therefore, a state-within-a-state.	257
What are Venezuela's major economic activities?	Lake Maracaibo oil, and iron ore in the east; oil reserves have also been discovered in the *llanos* regions of both Venezuela and Colombia.	258
What is the political status of each of the Guianas?	Guyana is independent, with unstable leadership; Suriname is also unstable, is now independent of the Dutch; French Guiana remains an overseas *department* of France	259-260

The Andean West

Which states make up this region? Which culture dominates? Which group rules?	Peru, Ecuador, Bolivia, and transitional Paraguay. The Amerind-subsistence cultural sphere is dominant, with a large *mestizo* population besides the majority Amerindians. Nonetheless, the European elite holds most of the political power.	260
What are the three physiographic/cultural subregions into which Peru is divided?	(1) The desert coast, the European-mestizo region; (2) the Andean Highlands or Sierra, the Amerindian region; (3) the eastern slopes and *montaña*, the sparsely populated Amerindian-mestizo interior.	260
What crucial territory did Bolivia lose to Chile?	Bolivia lost its Pacific outlet in the late 19th century, which leaves the country in a landlocked, disadvantageous situation.	264

The Southern Cone

In what ways does Paraguay exhibit traits of a regional transition zone?	Paraguay is associated with both the Andean West and the Southern Cone. About 95% of Paraguay's population is *mestizo*, with very strong Amerindian influences; Amerindian Guaraní is spoken alongside Spanish. Physiography is non-Andean, and Paraguayan economic geography is reorienting to the south.	265 - 266
Which states make up this region?	Argentina, Uruguay, and Chile. Paraguay continues to have strong links to this region.	266
What is the significance of the Argentine Pampa?	Productive meat-and-grain region which has made the country a major food exporter; gave Buenos Aires a rich hinterland, spurring its growth and success.	266-268
What is the economic situation in Argentina in the first decade of the 21st century?	Agriculture and manufacturing sectors experienced huge increases in the 1990s, but a severe recession and economic collapse occurred in the first ten years of the 21st century.	270
What are northern Chile's major minerals?	Nitrates in the Atacama Desert; copper, too, especially in the vicinity of Chuquicamata	271-272
Compare and contrast the territorial shapes of Uruguay and Chile.	Uruguay's is compact, easy to govern and manage; Chile's is severely elongated, causing potential political problems.	272

Brazil

How large is Brazil's relative size?	Territorially, it occupies just under 50% of South America; it ranks fifth in the world after Russia, Canada, the United States, and China	273
What centripetal forces bind Brazil together?	Successful ethnic mixing, adherence to the Catholic faith, a common language, and a cultural heritage of common music, fashion, and art. Commercial agriculture,	275

How has the Brazilian economy fared since 1980?	As of 2011, a more efficient subsidy program began implementation, with significant effects in the poorest States.	213
What is Brazil's politico-geographical framework?	A federal republic consisting of 26 States, and the federal district of Brasília.	214
What is the growth pole concept?	A development plan in which a set of "seed" industries are nurtured, thereafter setting off "ripples" of growth in the surrounding area.	216

MAP EXERCISES

Map Comparison

1. The impact of the Andes on South America's physical and cultural geography is discussed on pp. 199-205. Compare the maps referred to, setting forth your observations in some detail. Reread the section of chapter 4 on altitudinal zonation (p. 199), and discuss the application of this scheme to the Andes in the country of Peru.

2. South America's complex cultural geography often affects the ability of a state to integrate and effectively govern its citizens. After carefully studying Fig. 5A-4 (p. 241), rank the realm's countries according to their apparent cultural uniformity— assuming that the more homogeneous cultures are the stronger nation-states. Provide a brief justification for each country, but qualify it if you have learned appropriate additional information that might affect overall unity (or lack thereof).

3. On Fig. 5B-4 (text p. 261), rule a straight line in light pencil from Callao-Lima on the coast to Iquitos in the northeastern *Oriente*; draw an identical line on the physiographic map (Fig. 5A-1, p. 232). On a blank piece of paper, draw a cross-section of this traverse (using both maps to assist you), showing the general configuration of the surface involved (your instructor can assist you in doing this). Then, using Fig. 5A-4 (p. 241), add in the cultural association of each segment of the cross-section, and describe its leading characteristics according to the text discussion (pp. 239-240).

Map Construction   *(Use outline maps at the end of this chapter)*

1. In order to familiarize yourself with South American physical geography, place the following on the first outline map:

   a.   *Rivers*: Amazon, Orinoco, Paraguay, Paraná, Uruguay, São Francisco, Magdalena, Cauca, Apure, Marañon, Madeira, Guayas, Rio de la Plata, Colorado, Xingu, Tocantins, Négro

   b.   *Water bodies*: Mouth of the Amazon, Lake Maracaibo, Lake Titicaca, Gulf of Guayaquil, Strait of Magellan, Plata Estuary, Peru (Humboldt) Current

   c.   *Land bodies and features*: Cape Horn, Tierra del Fuego, Falkland Islands, Chiloé Island, Guiana Highlands, Cerrado, Andean Altiplano, Patagonian Plateau, Pampa, Chaco, Llanos, Brazilian Highlands, Atacama Desert, Andes Mountains

2. On the second map, political-cultural information should be entered as follows:

   a. Label each country, and its capital (*)

   b. Reproduce the cultural map (Fig. 5A-4, p. 241) using an appropriate color-pencil scheme

3. On the third outline map, urban-economic information should be entered as follows:

   a. *Cities* (locate and label with the symbol ●): Caracas, Ciudad Bolívar, Georgetown, Paramaribo, Cayenne, Valencia, Barquisimeto, Cartagena, Medellín, Bogotá, Buenaventura, Quito, Esmeraldas, Guayaquil, Lima, Callao, Iquitos, Huancayo, Arequipa, La Paz, Oruro, Potosí, Asunción, Arica, Antofagasta, Chuquicamata, Valparaíso, Santiago, Concepción, Valdívia, Montevideo, Buenos Aires, Rosario, Córdoba, Mendoza, Tucumán, Punta Arenas, São Paulo, Rio de Janeiro, Santos, Florianópolis, Pôrto Alegre, Belo Horizonte, Curitiba, Brasília, Volta Redonda, Salvador, Recife, Belém, Fortaleza, Manaus

   b. *Economic regions* (identify with circled letter):

      A - Itaipu Dam
      B - Cerrado
      C - Pampas
      D - *Oriente*
      E - Magdalena Valley
      F - Guayas Lowland
      G - Titicaca Basin
      H - Middle Chile
      I - Minas Gerais
      J - Grande Carajás Scheme

PRACTICE EXAMINATION

Short-Answer Questions

Multiple-Choice

1. South America's largest city in population size is:

   a) Mexico City b) São Paulo c) Buenos Aires
   d) Rio de Janeiro e) Caracas

2. The landlocked country whose quest for an outlet to the sea was thwarted by Chile is:

   a) Bolivia b) Suriname c) Portuguese Guiana
   d) Paraguay e) Amazonia

3. Which of the following regions is not located in Argentina?

   a) Patagonia b) Pampa c) Maracaibo Lowland
   d) Chaco e) Entre Rios

4. Brazil does not border:

   a) Paraguay b) Venezuela c) Chile
   d) Peru e) Argentina

5. Which of the following countries harbored an insurgent state at the opening of the twenty-first century?

   a) Uruguay b) Paraguay c) Brazil
   d) Colombia e) Argentina

6. Which of the following countries is a member of Mercosur?

   a) Argentina b) Bolivia c) Mexico
   d) Ecuador e) Suriname

True-False

1. Guyana is a former colony of France, and today enjoys complete independence.

2. Uruguay is a classic example of an elongated state.

145

3. Itaipu Dam is situated on the border between Brazil and Chile.

4. Belo Horizonte and Salvador are Amazon Basin growth poles.

5. Ecuador's largest city is also the country's capital.

6. Suriname is not one of the three Guianas.

Fill-Ins

1. The two leading resources of Chile's Atacama Desert are nitrates and _____.

2. At the center of the "Latin" American City model, one finds a land-use zone called the _____.

3. The coffee-growing areas of Colombia are concentrated in the altitudinal zone called *tierra* _____.

4. Brazil's largest city is _____.

5. Argentina's southern plateau region is called _____.

6. The country that borders Chile on its north is _____.

Matching Question on South American Countries

_____ 1. Still under colonial rule
_____ 2. Former Dutch Guiana
_____ 3. Pampa
_____ 4. Cuzco Basin
_____ 5. Contains realm's largest city
_____ 6. East side of Plata estuary
_____ 7. Elongated state
_____ 8. Capital is La Paz
_____ 9. Lower Orinoco Basin
_____ 10. Population more than 50% Asian
_____ 11. Site of Cusiana-Cupiagua oilfield
_____ 12. Regional transition zone
_____ 13. Guayas Lowland

A. Colombia
B. Venezuela
C. Guyana
D. Suriname
E. French Guiana
F. Ecuador
G. Peru
H. Bolivia
I. Paraguay
J. Chile
K. Argentina
L. Uruguay
M. Brazil

Essay Questions

1. Now overcoming internal social and economic problems, BRIC-member Brazil possesses vast growth potential. Discuss the advantages of Brazil's developmental opportunities, emphasizing its natural resources in a subregion-by-subregion survey.

2. Discuss the altitudinal zonation scheme of Middle and South American environments (introduced on text p. 199 in Chapter 4) as it applies to the Andean highlands of northwestern South America, highlighting the distinctive cultural and economic geographies associated with each zone.

3. Discuss the historical geography of Peru over the past 1000 years. Why did great advances occur here in the pre-European period? Why did the Spaniards headquarter their colonial empire here after 1535? Why has Peru ceased to be one of South America's leading countries in the past century?

4. Discuss the agricultural geography of South America. Draw a sketch map showing the commercial, subsistence, and non-agricultural areas of the realm. Why did this particular pattern emerge, and what relationship does it bear to the cultural- spatial differentiation of the continent?

5. The economic geography of northern and western South America has in part been shaped by the exploitation of major mineral resources. Discuss the spatial patterns of production that characterize the following metallic and fossil-fuel resources: oil and natural gas, nitrates, copper, silver, and iron ore. What key minerals are missing or are unfavorably distributed in these countries?

SOUTH AMERICA

| 0 | 400 | 800 | 1200 | 1600 Kilometers |
| 0 | 200 | 400 | 600 | 800 | 1000 Miles |

Equator

Tropic of Capricorn

SOUTH AMERICA

0   400   800   1200   1600 Kilometers
0 200 400 600 800 1000 Miles

Equator

Tropic of Capricorn

SOUTH AMERICA

CHAPTERS 6 A&B
SUBSAHARAN AFRICA

OBJECTIVES OF THESE CHAPTERS

Chapters 6 A and B treat Subsaharan Africa, a problem-plagued realm in which life for the masses is almost always difficult. Following the introduction, Africa's environmental base is covered, highlighting physiography and disease patterns. Historical geography comes next, examining pre-European cultures, the colonial transformation after 1450, and the post-independence era that dates from the late 1950s. Agricultural potential (as well as risks) is covered, and the regional overview focuses in turn on Southern, East, Equatorial, and West Africa as well as the African Transition Zone.

Having learned the regional geography of Subsaharan Africa, you should be able to:

1. Understand the interpretation of African physiography and hydrography that is offered by the continental drift hypothesis.

2. Appreciate the concerns of contemporary medical geography and their applications to Africa.

3. Understand the course of Subsaharan African history from the indigenous cultures through the colonial and post-independence eras.

4. Explain the political territorialization of modern Africa as a legacy of the colonial era that has finally ended.

5. Grasp the importance of agriculture in African economic life, and the severe environmental risks that farming and herding are exposed to.

6. Understand the cultural and economic trends that have shaped the regional structuring of Southern, East, Equatorial, and West Africa.

7. Understand the present politico-geographical situation in South Africa, including the racial polarization that intensified under the now-dismantled *apartheid* system, and the country's subsequent shift to democracy.

8. Locate the leading physical, cultural, and economic-spatial features of the realm on an outline map.

GLOSSARY

Chapter 6A

Human Evolution (284)

The long term biological maturation of the human species. Geogrpahically, all evidence points toward East Africa as the source of humankind. *Homo sapiens,* emigrated from this hearth to eventually populate the rest of the ecumene.

Rift valleys (285)

Geologic trenches formed when huge parallel cracks or *faults* occur in the Earth's crust, causing in-between strips of land to sink and form extensive linear valleys.

Continental drift (285-86)

The hypothesis underlying the formation of Africa's physiography and hydrography. The breakup of the supercontinent called Pangaea and the subsequent drifting apart of its landmass components; Africa occupied the heart of Gondwana (Pangaea's southern part), and its physical geography is still shaped by this geologic functioning as the core-shield.

Dhows (287)

Wooden boats with triangular sails, plying the seas of the Indian Ocean between the Arabian and the East African coasts.

State Formation (290)

The creation of a state based on traditions of territoriality that go back thousands of years.

Berlin Conference (292; 296-box)

The 1884-1885 international conclave, convened by Germany's Chancellor von Bismarck, which superimposed on Africa a system of political boundaries agreed to by all colonial powers; amazingly, the scheme has survived for over a century and this political fragmentation, for the most part, still underpins the realm's contemporary politico-geographical structuring.

Multilingualism (302-303)

A society marked by a mosaic of many local languages. Constitutes a centrifugal force because it impedes communication with the larger population. Often a lingua franca is used as a common language in many countries of Subsaharan Africa.

Formal Sector (304)

The total activities of a country's legal economy that is taxed and monitored by the government, whose **gross domestic product (GDP)** and **gross national product (GNP)** are based on it; as opposed to an **informal economy**.

Informal Sector (304)

Dominated by unlicensed sellers of homemade goods and services, the primitive form of capitalism found in many developing countries that takes place beyond the control of government. The complement to a country's **formal sector.**

Land tenure (297)

The way that people own, occupy, and use land.

Land alienation (297)

Expropriation of (often the best) land by a conquering group.

Colonialism (297-298)

The drive toward the creation and expansion of a colonial empire and, once established, its perpetuation.

Green Revolution (299)

The development of higher-yield, fast-growing varieties of rice and other cereals in certain developing countries. This led to increased production per unit area and a temporary narrowing of the gap between population growth and food needs; today, unfortunately, that gap is widening again.

Medical geography (300)

The subdiscipline concerned with the spatial aspects of health and illness.

Endemic (300)

Refers to a disease in a host population that affects many people in a kind of equilibrium without causing rapid and widespread deaths.

Epidemic (300)

A local or regional outbreak of a disease.

Pandemic (300)

An outbreak of a disease that spreads worldwide.

Chapter 6B

Apartheid (312)

Literally, apartness. The Afrikaans term for South Africa's pre-1994 policies of racial separation, a system that produced highly segregated socio-geographical patterns.

Separate development (312-313)

The spatial expression of South Africa's "grand" apartheid scheme, whereby nonwhite groups were required to settle in segregated "homelands." The policy was dismantled when white-minority rule collapsed in the early 1990s.

Highveld (313)

Dutch term for the high plateau of the South African interior.

Afrikaners (313)

The descendants of the original Dutch colonists of South Africa, known as the Boers, who steadily gained numerical and political strength until they came to power as the white-minority government from 1948-1994.

Coloured people (313)

Term for the racially mixed population resulting from the intermarriage of European settlers and Africans in South Africa's Cape area. Today, this mixed ancestry group forms the majority population in and around Cape Town.

Exclave (319)

A bounded (non-island) piece of territory that is part of a particular state but lies separated from it by the territory of another state.

Landlocked state (323)

A state with no outlet to the sea. Examples in this realm are Ethiopia, Burkina Faso, Chad, Uganda, and Zimbabwe.

Sharia Law (331-332)

Strict Islamic law.

Enclave (335)

A piece of territory that is surrounded by another political unit of which it is not a part.

Periodic market (335)

Village market that opens once every few days; part of a regional network of similar markets in rural, preindustrial societies where goods are brought to market on foot and barter remains a major mode of exchange.

Islamic Front (335-336)

The southern border of the African Transition Zone that marks the frontier of the Muslim faith in its southward penetration of Subsaharan Africa.

Choke Point (336)

A narrowing of an international waterway causing marine traffic congestion, requiring reduced speeds and/or sharp turns, and increasing the risk of collision as well as vulnerability to attack. When the waterway narrows to a distance of less than 38 kilometers (24 mi), this necessitates the drawing of a **median line (maritime) boundary**. Examples are the Hormuz Strait between Oman and Iran at the entrance to the Persian Gulf, and the Strait of Malacca between Malaysia and Indonesia.

Failed State (338)

A country whose institutions have collapsed and in which anarchy prevails. Somalia is a current example.

SELF-TESTING QUESTIONS

Cover the right side of the page with a sheet of paper. Uncover each line after you have attempted to answer the question in the left column. If necessary, refer to textbook page(s) listed at the right.

Question	Answer	Page
African Environments		
What is a rift valley?	A trench formed when cracks or *faults* occur in the Earth's crust, and in-between strips sink down to create linear valleys.	285
What is *continental drift*? How does it help us to understand Africa's physiography?	The hypothesis that all continents were once joined into a single supercontinent (Pangaea); Africa was located centrally in the southern component (Gondwana); the giant landmass began to break apart at least 200 million years ago, with its continental components "drifting" away toward their present locations. Africa's unusual topographic and drainage patterns are most readily explained by this hypothesis.	285-286
Describe Africa's broad patterns of climate and vegetation.	They are symmetrically distributed about the equator, which bisects the continent. The humid tropics are largely restricted to the western half since East Africa is mainly an elevated plateau; as one moves poleward from the tropics, steppes, significant deserts, and narrow strips of Mediterranean climates successfully prevail.	286; 294
African Agriculture And Environment		
What problems plague African farming?	Poor tropical soils, excessive droughts, soil exhaustion, inadequate equipment, lack of capital, and inefficient methods.	297
What is the prospect for agricultural improvement?	Bleak for the foreseeable future; in the post-colonial period, food production has stagnated and at times declined; the life of the farmer remains difficult and gains in African agricultural output overall are more than offset by the rapid rate of population growth.	297

| How do natives of the Sahel region cope with their harsh environment? | They herd livestock and diversify their agricultural crops for the varying amounts of moisture available throughout the year. | 297; 335 |

Medical Geography

What is *medical geography*?	The study of health and disease within a geographic context and from a spatial perspective. Among other things, this field of geography examines the sources, diffusion routes, and distributions of diseases.	300
What is the difference between an epidemic and a pandemic disease?	Epidemics are regional phenomena; an outbreak of global proportions is described as a pandemic.	300
What is Africa's worst pandemic disease?	Africa and the world's deadliest vectored disease is malaria, whose carrier is a mosquito that prevails in most of Africa's inhabited areas.	300

Historical Geography

Why is so little known about Subsaharan Africa before A.D. 1500?	The absence of written histories is the main cause, exacerbated by neglect and destruction of traditions and artifacts following the onset of colonialism and the disruptions of the slave trade.	287-288
Does this mean that pre-colonial Africa had no organized cultures and traditions?	Definitely not; on the contrary, sophisticated artifacts throughout Africa suggest advanced societies, widespread trade, and rich cultural legacies.	287-288
Where did the first European contacts occur?	In the 15th century, Europe's ships began to ply the waters of the entire west coast of Africa as the route to southern Asia opened. Way-station ports opened, and soon African middlemen began to organize trading of slaves, minerals, ivory, and spices.	290

How did the subsequent European penetration proceed?	Slowly at first, but after 1800 European powers finally laid claim to all of Africa's territory; expansion of spheres of influence soon caused disagreements but the 1884-85 Berlin Conference solved things peaceably.	292
What administrative systems did the Europeans employ in colonial Africa?	Indirect rule of colonies, protectorates, and territories (Britain); paternalism (Belgium); assimilation toward the mother country (France); rigid economic control (Portugal).	292; 294-box

Cultural Patterns

How many languages are spoken in this realm?	Over one thousand—dozens of languages in single countries, hundreds in a single region—make up the intricate linguistic jigsaw of Subsaharan Africa.	301
Which religion was brought to Subsaharan Africa with the first great wave of proselytism?	Islam.	303

East Africa

Describe East Africa's natural environment.	Plateau savanna country that becomes a steppe in the drier northeast; volcanoes and rift valleys mark the surface; Lake Victoria is a key feature.	319
Which countries constitute this region?	Tanzania, Kenya, Uganda, Rwanda, Burundi, and highland (southwestern) Ethiopia. See Fig. 6B-4, p. 320.	320-321
How does Kenya's development differ spatially from Tanzania's?	Kenya's development is concentrated in the core area around Nairobi; Tanzania's is much more dispersed.	322
What has Tanzania's course been since independence?	It embarked on a socialist development program, but without adequate planning, so results were disappointing. Prospects have improved: a market-oriented "recovery" is now being employed, and the tourism industry is on the rise.	322-323

What is Uganda's current status?	The difficult struggle to emerge from the chaotic Amin years of the 1970s still continues. Economic recovery has been painstakingly slow; there has been some gains against the AIDS pandemic. Political conflict continues.	323

Equatorial Africa

Which countries make up this region?	The Congo (former Zaire), Cameroon, Central African Republic, Congo, Gabon, Equatorial Guinea, southern Chad, and the southern subregion of Sudan. See Fig 6B-6, p. 326.	324-326
What has hampered The Congo's great development potential?	Lack of national cohesion, corrupt rule, and infrastructure problems.	325

West Africa

Which countries are included in this region?	Mauritania, Mali, Niger, Senegal, Gambia, Guinea-Bissau, Guinea, Sierra Leone, Liberia, Burkina Faso, Ivory Coast, Ghana, Togo, Benin, Nigeria, and part of Chad could be included too. See Fig. 6B-7, p. 329.	328-329
How is Nigeria's government structured?	It is a federal system consisting of 36 states.	330
Which was the first of West Africa's states to become independent?	Ghana, in 1957.	332
What are *periodic markets?*	Local village markets that operate at regular intervals every few days; they attract nearby buyers and sellers.	335

African Transition Zone

What is the extent of the region?	The east-west band of states along the southern margin of the Sahara, from Mauritania and Senegal in the west to Ethiopia and Somalia in the eastern Africa Horn, and then southward to include the narrow coastal strips of Kenya and Tanzania. See Fig 6B-9, p. 336.	335
Where is the Islamic Front?	The Islamic Front forms the southern border of the African Transition Zone in its west-to-east component from Sierra Leone to northern Ethiopia; from there southward it forms the region's western border in eastern Ethiopia, southwestern Somalia, and coastal Kenya and Tanzania.	335-336

MAP EXERCISES

Map Comparison

1. Many of Africa's countries contain diverse cultural groups that were largely thrown together by the Europeans in the political territories that were carved out during the colonial era. Compare the maps on pp. 289, 290, 293, and 302.. Record your observations about this fragmented pattern, its evolution since the colonial era, the countries that are most strongly affected, and how recent population movements in the post-independence era are adjusting some established cultural-geographic patterns.

2. Compare and contrast the map on p. 291 with the map in Figure A in Appendix A. Why did the trans-Atlantic slave trade become such a lucrative operation, achieving mass proportions in such a short time period? Why did West Africa spawn so prominent a slave trade? How does the present social geography of the Caribbean and northeastern South America reflect the aftermath of the Atlantic slave trade (refer back to appropriate sections of Chapters 4 and 5)?

Map Construction (*Use outline maps at the end of this chapter*)

1. In order to familiarize yourself with African physical geography, place the following on the first outline map:

   a. *Rivers*: Niger, Senegal, Volta, Benue, Congo, Lualaba, Ubangi, Kasai, Blue Nile, White Nile, Zambezi, Cubango, Limpopo, Vaal, Orange

   b. *Water bodies*: Gulf of Guinea, Moçambique Channel, Lake Chad, Lake Volta, Lake Victoria, Lake Turkana, Lake Tanganyika, Lake Albert, Lake Malawi, Lake Kariba

   c. *Land areas:* Madagascar, Futa Jallon Highlands, Ethiopian Highlands, Drakensberg, the Sahara, the Sahel, Kalahari Desert, Namib Desert, Adamawa Massif, Bihe Plateau, Cape of Good Hope

2. On the second map, political information should be entered as follows:

   a. Label each country with its name.

   b. Locate and label each capital city with the symbol *.

3. On the third outline map, urban-economic information should be entered as follows:

a. *Cities* (locate and label with the symbol ●): Dakar, Bamako, Niamey, Conakry, Freetown, Monrovia, Ouagadougou, Abidjan, Kumasi, Accra, Lomé, Porto Novo, Kano, Ibadan, Lagos, Enugu, Port Harcourt, Maiduguri, N'Djamena, Adis Abeba, Nairobi, Mombasa, Kisumu, Kampala, Jinja, Juba, Dar-es-Salaam, Mwanza, Yaoundé, Douala, Matadi, Kinshasa, Bangui, Brazzaville, Kananga, Lubumbashi, Luanda, Lobito, Lusaka, Ndola, Blantyre, Harare, Bulawayo, Beira, Maputo, Antananarivo, Johannesburg, Pretoria, Cape Town, Bloemfontein, Durban, Port Elizabeth, East London

b. *Economic regions* (identify with circled letter):

    A - Copperbelt
    B - Great Dyke
    C - Witwatersrand
    D - Niger Delta
    E - Great Lakes region
    F - Transvaal
    G - Highveld

PRACTICE EXAMINATION

Short-Answer Questions

Multiple-Choice

1. The edge of the African plateau, especially in Southern Africa, is known as the:

    a) Rift Valley   b) Gondwana Mountains   c) Great Dyke
    d) Islamic Front   e) Great Escarpment

2. The West African country that shifted its capital from Lagos to the centrally-located city of Abuja is:

    a) Nigeria   b) Tanzania   c) Sudan
    d) Burkina Faso   e) Ivory Coast

3. South Africa's population of mixed European-African ancestry that resides in the Cape area is known as the:

    a) Coloured sector   b) Durbans   c) Afrikaners
    d) Zulu   e) Indians

4. Nelson Mandela was the leader of:

    a) Chad   b) Equatorial Africa   c) Tanzania
    d) the Sahel   e) South Africa

5. Which of the following countries borders Lake Victoria?

    a) Uganda   b) Nigeria   c) Zambia
    d) South Sudan   e) Somalia

6. The first Europeans to colonize what is now South Africa come from:

    a) Britain   b) France   c) Belgium
    d) the Netherlands   e) Germany

True-False

1. Before independence, the modern state of The Congo was a colony of Britain.

2. Nigeria's government is organized as a federal system.

3. The Copperbelt is partially located in the country formerly known as the Belgian Congo.

4. Africa is known as the "plateau continent."

5. East Africa lies at a lower general elevation than Equatorial Africa.

6. Johannesburg is South Africa's largest seaport.

7. The region known as the African Transition Zone contains major desert areas.

Fill-Ins

1. The Ibo and Yoruba peoples inhabit the country called _____.

2. The racial policy of South Africa was known as _____, the successor of apartheid before both systems were dismantled in the mid-1990s.

3. The worst and most widespread of Africa's pandemic diseases is _____.

4. South Africa's co-capitals are Pretoria and _____.

5. The country formerly called Zaïre is today known as _____.

6. The large island lying to the east of Southern Africa is _____.

Matching Question on African Countries

____1. Hausa-Fulani	A. Tanzania
____2. Former Portuguese colony	B. The Congo
____3. Former French Colony	C. Equatorial Guinea
____4. Boers	D. Zimbabwe
____5. Contains the Copperbelt	E. Moçambique
____6. Dar-es-Salaam	F. Nigeria
____7. Great Dyke	G. Senegal
____8. Island of Bioko	H. South Africa
____9. Country named after a desert	I. Zambia
___10. Former colony of Belgium	J. Namibia

Essay Questions

1. Discuss at some length the concerns of contemporary medical geography, particularly as they apply to Subsaharan Africa today.

2. The continental drift hypothesis contributes a great deal to the explanation of African hydrography (surface water patterns) and physiography. Show how one can use this approach to understand Africa's dominantly plateau-like terrain, the absence of mountain chains, the rift valleys, and the unusual courses of the realm's major rivers.

3. Perhaps no region has been more thoroughly disrupted and transformed by the colonial experience than West Africa. Discuss the major European penetrations here, the kinds of political territories that emerged, and how some of the region's disparate populations came into conflict as Islam has attempted to carve a stronger foothold in Nigeria.

4. The possibility of a "Green Revolution" for Africa is widely debated today. Discuss the present condition of Subsaharan Africa's agriculture, the nature of the Green Revolution itself, and chances for success if it were widely implemented in this realm.

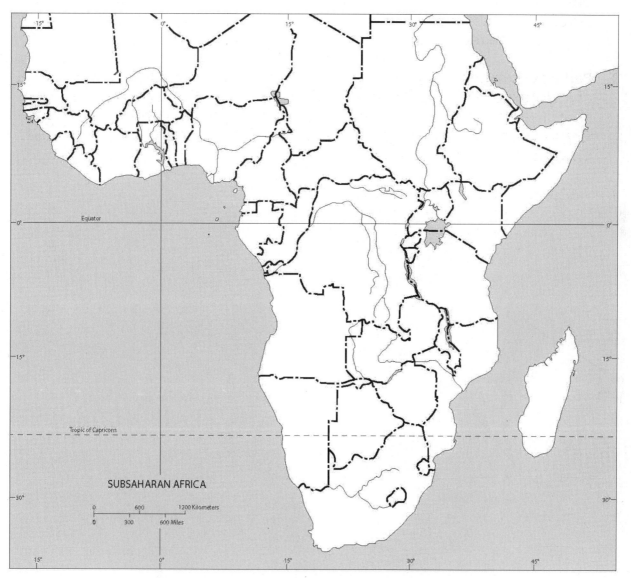

SUBSAHARAN AFRICA

Equator

Tropic of Capricorn

0   600   1200 Kilometers
0   300   600 Miles

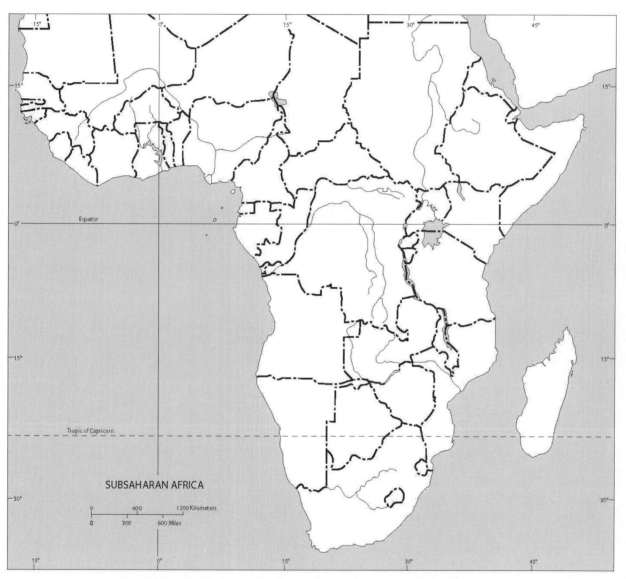

SUBSAHARAN AFRICA

Equator

Tropic of Capricorn

15°   0°   15°   30°   45°

15°

0°

15°

30°

0   600   1200 Kilometers

0   300   600 Miles

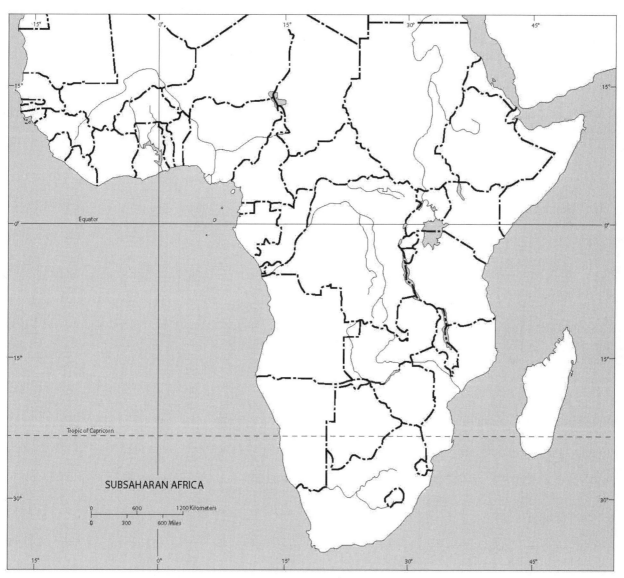

Equator

Tropic of Capricorn

SUBSAHARAN AFRICA

| 0 | | 600 | | 1200 Kilometers |
| 0 | 300 | | 600 Miles | |

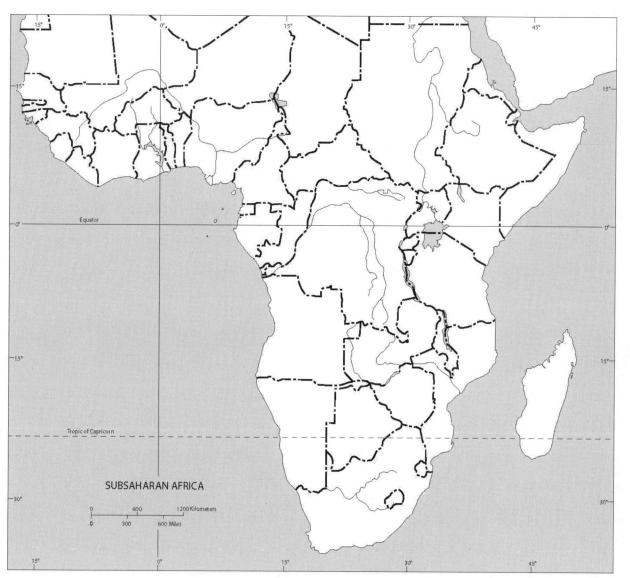

SUBSAHARAN AFRICA

Equator

Tropic of Capricorn

0   600   1200 Kilometers
0   300   600 Miles

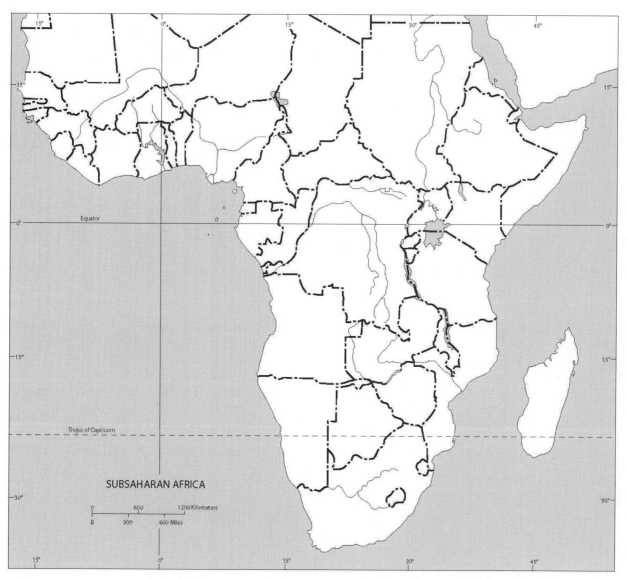

SUBSAHARAN AFRICA

0   600   1200 Kilometers

0   300   600 Miles

CHAPTERS 7A&B
NORTH AFRICA / SOUTHWEST ASIA

OBJECTIVES OF THESE CHAPTERS

Historically, the North Africa/Southwest Asia realm is properly regarded as the world culture hearth. Following a brief introduction and discourse on defining this complex realm, its historical evolution is traced emphasizing the prominence of religion, particularly Islam. Traditional culture, of course, stills weighs heavily on the daily affairs of this realm, and even its vast oil production is influenced by these considerations. An in-depth regional treatment follows, highlighting such major trouble spots as Israel, Iraq, Iran, Sudan, Cyprus, Lebanon, the Horn of Africa, and the Persian Gulf. Turkestan is the realm's seventh region; this is former Soviet Central Asia, (now constituted by Kazakhstan, Turkmenistan, Uzbekistan, Kyrgyzstan, Tajikistan), to which we have added Afghanistan.

Having learned the regional geography of North Africa/Southwest Asia, you should be able to:

1. Appreciate the complexities involved in defining and naming this realm.

2. Understand the realm's basic cultural geography.

3. Describe the history of this realm, stressing its role in the development of many of the world's leading religions, particularly Islam.

4. Appreciate the significance of Islam for this realm as a whole, and the internal geographic variations of that faith.

5. Explain the major processes of spatial diffusion and be aware of the broad geographic patterns they shape.

6. Describe the production of oil in this realm, and the impact it has had on the development of countries that contain petroleum supplies.

7. Understand the major trends within each of this realm's regions, and why so many global political problems have arisen here.

8. Locate the major physical, cultural, and economic-spatial features of the realm on an outline map.

Cultural geography (345)

The wide-ranging and comprehensive field of geography that studies spatial aspects of human cultures.

Culture hearth (345)

A source area or innovation center from which cultural traditions are transmitted.

Cultural diffusion (345)

The outward spreading of a culture trait from its hearth to other places.

Cultural landscape (345)

The forms and artifacts sequentially placed on the natural landscape by the activities of various human occupants. By this progressive imprinting of the human presence, the physical (natural) landscape is modified into the cultural landscape, forming an interacting unity between the two.

Mesopotamia (345-346)

The Tigris-Euphrates Plain of present-day Iraq—literally "land amidst the rivers"—which is the hearth of civilization.

Fertile Crescent (346)

An arc stretching from the eastern Mediterranean coast to near the Persian Gulf, site of early plant domestications and farming innovations (see Fig. 7-3).

Hydraulic civilization theory (346)

Civilizations able to control irrigated farming over large hinterlands; often held power over others in less fortuitous locations.

Climate change theory (347)

An alternative to the *hydraulic civilization theory*; holds that changing climate (rather than a monopoly over irrigation methods) could have provided certain cities in the ancient *Fertile Crescent* with advantages over others.

Spatial diffusion (349)

The spatial spreading or dissemination of a phenomenon across space and through time.

Expansion diffusion (349)

The spreading of an idea or innovation through a fixed population in such a way that the number of those adopting grows continuously larger.

Relocation diffusion (349)

Diffusion by migration wherein innovations are carried by a relocating population.

Contagious diffusion (349)

Local-scale diffusion, strongly controlled by distance from the point of origin.

Hierarchical diffusion (349)

Macro-scale diffusion through a national or continental-scale urban hierarchy, involving the "trickling down" of an innovation from atop a hierarchy to each of the lower levels in turn.

Islamization (349)

Introduction and establishment of the Muslim religion. A process still underway, most notably along the Islamic Front, that marks the southern border of the African Transition Zone.

Stateless nation (354; 376; 380)

A national group that aspires to become a nation-state but lacks the territorial means to do so; the Palestinians and Kurds of Southwest Asia are classic examples.

Choke point (355; 387)

A narrowing of an international waterway causing marine traffic congestion, requiring reduced speeds and/or sharp turns, and increasing the risk of collision as well as vulnerability to attack, such as the Hormuz Strait at the entrance to the Persian Gulf.

Fragmented Modernization (358)

A checkerboard-like spatial pattern of **modernization** in an **emerging-market** economy wherein a few localized regions of a country experience most of the development while the rest are largely unaffected.

Religious fundamentalism (revivalism) (359)

Religious movement whose objectives are to return to the foundations of that faith and to influence state policy.

Wahhabism (360)

A particularly virulent form of (Sunni) Muslim revivalism that was made the official faith when the modern state of Saudi Arabia was founded in 1932.

Jihad (360)

A doctrine within Islam. Commonly translated as *holy war*, it entails a personal or collective struggle on the part of **Muslims** to live up to the religious standards prescribed by the *Quran* (Koran).

Domino effect (361)

The belief that political destabilization in one **state** can result in the collapse of order in a neighboring state, triggering a chain of events that, in turn, can affect a series of **contiguous** states.

Cultural revival (366)

The regeneration of a long-dormant culture through internal renewal and external infusion.

Basin irrigation (367)

An ancient irrigation method of the lower Nile Valley involving the trapping and later release of floodwaters.

Perennial irrigation (367)

The more modern Egyptian irrigation technique, using dams and levees to store and regulate the use of floodwater throughout the year.

Fellaheen (369)

Egypt's peasant farmers, who still struggle to eke out a subsistence level of existence.

Maghreb (372)

The western region of this realm, consisting of the northwesternmost African countries of Morocco, Algeria, and Tunisia. The name itself means "Isle of the West."

Tell (373)

The lower slopes and coastal plains of northwesternmost Africa between the Atlas Mountains and the sea.

Qanats (392)

Underground tunnels that carry water to dry flatland areas.

Buffer state (399)

Part of a *buffer zone*–a set of countries separating ideological or political adversaries. In southern Asia, Afghanistan, Nepal, and Bhutan were parts of a buffer zone between British and Russian-Chinese imperial spheres. Thailand was a *buffer state* between British and French colonial domains in mainland Southeast Asia.

SELF-TESTING QUESTIONS

Cover the right side of the page with a sheet of paper. Uncover each line after you have attempted to answer the question in the left column. If necessary, refer to the textbook page(s) listed at the right.

Question	Answer	Page
Realm Characteristics		
Why is "Arab World" a misleading title for this realm?	This is a *linguistic* term which does not fit much of the realm.	344
Why is "Islamic World" just as unsatisfactory?	The Muslim religion prevails far beyond the realm, and within it there are a number of countries in which Islam is not the dominant faith.	344-345
Why is "Middle East" not much of an improvement?	It is imprecise and reflects a Western perceptual bias—but it is based on multiple criteria (unlike the two preceding labels).	344
Cultural Geography		
What is cultural geography?	The wide-ranging and comprehensive field that studies spatial aspects human cultures, focusing on not only culture landscapes but also culture hearths.	345
What are the goals of those who specialize in cultural diffusion?	To be able to reconstruct ancient routes by which knowledge and achievements of culture hearths spread (diffused) to other areas.	345
Historical Geography		
What were the three culture hearths of North Africa/South-west Asia?	The Tigris-Euphrates Plain (Mesopotamia), the lower Nile Valley, and the peripheral lower Indus Valley (which in this book is treated as part of the South Asian realm).	346
Name some of the major agricultural innovations that originated in these hearths.	Wheat, rye, barley, peas, beans, grapes, apples, peaches, horses, pigs, and sheep.	346

| When and where did the Prophet Muhammad live? | A.D. 571-632; Muhammad lived in Mecca, a town in the Jabal Mountains. | 348 |

Spatial Diffusion

What is an *innovation wave*?	As explained by Hägerstrand's model, the progress of a diffusing phenomenon, which pulses outward from its hearth across the adopting region.	349
What are the two major types of spatial diffusion?	*Relocation diffusion* (diffusion by migration) and *expansion diffusion* (diffusion within a fixed population).	349
What are the two forms of expansion diffusion?	*Contagious diffusion* (distance-controlled local diffusion) and *hierarchical diffusion* (coursing downward within a national urban hierarchy).	349

Oil Resources

| What proportion of the world's oil reserves are contained within this realm? | About 57 percent. | 353 |
| What are the Big Five oil-producing nations? | Saudi Arabia, Iran, Iraq, United Arab Emirates, and Kuwait. | 353 |

Egypt and the Lower Nile Basin

Why is Egypt so prominent an Arab country?	Its pivotal location, maintained through history; it lies at the crossroads where North Africa and Southwest Asia come together–the heart of the Arab World.	367
How important is the Nile to this country?	This river is Egypt's lifeline; valley and delta are home to about 95% of the population.	367
How do *basin* and *perennial* irrigation differ?	The former is an ancient method of trapping floodwaters; the latter is a modern system based on dams, of which the Aswan High Dam is most prominent.	367-368

Name Egypt's six major subregions.	Nile Delta; Middle Egypt (including Cairo); Upper Egypt; Western Desert; Eastern Desert/Red Sea Coast; Sinai Peninsula. See Fig. 7B-2, p. 368.	367; 368-map

The Maghreb and Libya

What does Maghreb mean? Name its constituent countries.	Literally "Isle of the West," based on the Atlas Mountains standing above the flat Sahara. Morocco, Algeria, and Tunisia; neighboring Libya is often discussed together with the Maghreb.	372-373
Which country of the Maghreb is the most Westernized?	Tunisia.	372

The Middle East

Name the 5 countries of this region.	Iraq, Syria, Jordan, Lebanon, and Israel. See Fig. 7B-4, p. 377.	377-map
What are the leading problems of Iraq, Syria, and Jordan?	Iraq faces recovery after the 2003 invasion, and its infrastructure and government must be reconstituted; Syria's negotiations with Israel for the return of the Golan Heights have stalled; Jordan is beset by a rapidly growing population in which refugees outnumber natives, and filling the leadership void left by its late King Hussein.	376-379
What is the current situation in Lebanon?	After the Hizbollah kidnapping of Isreali soldiers and subsequent Israeli attack of 2006, its political and physical infrastructure remains shattered.	380
How old is the modern state of Israel?	It was founded amid much turmoil in 1948, and has lived with threatening neighbors ever since.	380
What is the distribution of Palestinians in the Middle East?	As of 2012, almost 12 million are projected throughout the region; Jordan contains 2.8 million, and Syria and Lebanon contain sizeable concentrations as well; Israel and its occupied territories are home to more than 5 million.	381

Arabian Peninsula

Name the countries of this region.	Saudi Arabia, Kuwait, Bahrain, Qatar, the United Arab Emirates, the Sultanate of Oman, and the Republic of Yemen. See Fig. 7B-9.	385-map
What are some major ongoing Saudi Arabian development projects?	The construction of an effective internal transportation and communications network; the growth of the petrochemical and metal industries; the cities of Jubail and Yanbu.	385-386

The Empire States

Name the countries of this region.	Turkey, Azerbaijan, Iran, and the northern part of Cyprus.	388
Why has Turkey remained aloof from other Arab countries?	Greater European influences shaped culture here; Atatürk's revolution stressed the detachment from other Islamic states, and that tradition endures.	390-391
How has modernization affected Iran?	Surprisingly little; it was also a contributory factor to the 1979 revolution that deposed the Shah.	392
How did the Iran-Iraq War (1980-1990) affect Iran?	The ten-year struggle left Iran poorer and weaker: oil revenues were spent on unproductive pursuits.	393

Turkestan

Which countries comprise Turkestan?	Kazakhstan, Turkmenistan, Uzbekistan, Kyrgyzstan, Tajikistan, and Afghanistan. Before 1992, these countries constituted the former Soviet Union's Central Asian republics (with the exception of Afghanistan). See Fig. 7B-12.	394; 395-map
Why is Kazakhstan's relative location important?	It forms a corridor between the Caspian oil reserves and China (a large oil consumer). Future pipelines could be crucial to China.	397

What problems plague Uzbekistan?	Contaminated groundwater from overuse of pesticides; severe medical problems resulting from this contamination; social upheaval from *Wahhabism*, a virulent form of Islamic fundamentalism.	397
What centrifugal forces divide Tajikistan?	The northern region of the nation is an area of anti-Tajik, Uzbek activism.	399
Which states established the buffer role for Afghanistan?	19th century Britain and Russia, for which the country was a neutral zone lying between their Asian spheres of influence.	399
How has Turkestan fared since the end of Soviet rule?	Government authoritarianism has continued–corruption, inefficiency, and instability plague this region; Islamic fundamentalism is on the rise; and the area suffers from ethnic and cultural strife.	399-401

MAP EXERCISES

Map Comparison

1. Compare the map of world religions (see Introductory chapter) to the map of world geographic realms (pp. 2-3 – intro chapter). Which realms are clearly identifiable with a major religion? Which realms exhibit multiple religions that suggest internal cultural conflicts? Double-check your observations against the world population map (pp. 20-21), underscoring those realms associated with large population clusters.

2. Prepare a brief essay on the distribution of petroleum reserves in the North Africa/Southwest Asia realm based on your analysis of Fig. 7A-8 (p. 356). When possible, compare to patterns of production as shown in the bar graph at the bottom of the map.

Map Construction *(Use outline maps at the end of this chapter)*

1. In order to familiarize yourself with North Africa/Southwest Asia physical geography, place the following on the first outline map:

    a. *Rivers*: Nile, White Nile, Blue Nile, Tigris, Euphrates, Orontes, Jordan, Shatt al Arab, Amu Darya, Syr Darya

    b. *Water bodies*: Red Sea, Persian Gulf, Suez Canal, Gulf of Aqaba, Gulf of Suez, Bab el Mandeb Strait, Hormuz Strait, Strait of Gibraltar, Gulf of Sidra, Caspian Sea, Aral Sea, Black Sea, Mediterranean Sea, Lake Nasser, Lake Tana, the Sudd, Dead Sea, Bosphorus Strait, Gulf of Adan, Gulf of Oman, Arabian Sea, Garagum Canal, and the new Mesopotamian Drainage Canal

    c. *Land bodies*: Sinai Peninsula, Cyprus, Horn of Africa, Socotra, Bahrain, Canary Islands, Kyrgyz (Kirghiz) Steppe

    d. *Mountains and Deserts*: Atlas Mountains, Tibesti Mountains, Libyan Desert, Sahara Desert, Arabian Desert, Rub al Khali (Empty Quarter), Anatolian Plateau, Ahaggar Mountains, Ethiopian Highlands, Hejaz Mountains, Negev Desert, Plateau of Iran, Elburz Mountains, Zagros Mountains, Pamir Mountains, Tian Shan Mountains, Hindu Kush Mountains, Nubian Desert, An Nafud Desert

2. On the second map political-cultural information should be entered as follows:

    a. Label each country, and label and locate each capital city with the symbol *.

b. Color each country according to its major religion, using information and appropriate color symbols from the world religions map in the introductory chapter (be sure to differentiate between the Sunni and Shi'ite sects in Muslim states).

3. On the third outline map, economic-urban information should be entered as follows:

a. *Cities* (locate and label with the symbol ● /capitals should again be shown with symbol *): Casablanca, Marrakech, Rabat, Tangiers, Oran, Algiers, Tunis, Tripoli, Benghazi, N'Djamena, Niamey, Bamako, Timbuktu (Tombouctou), Nouakchott, Port Sudan, Khartoum, Juba, Asmara, Djibouti, Mogadishu, Cairo, Alexandria, Port Said, Ismailia, Aswan, Riyadh, Mecca, Medina, Dhahran, Jubail, Yanbu, Kuwait City, Manama, Doha, Abu Dhabi, Muscat, Adan, San'a, Amman, Jerusalem, Haifa, Tel Aviv-Jaffa, Eilat, Aqaba, Beirut, Sidon, Damascus, Aleppo, Nicosia, Baghdad, Basra, Kirkuk, Mosul, Istanbul, Ankara, Izmir, Adana, Zonguldak, Diyarbakir, Tehran, Tabriz, Abadan, Mashhad, Shiraz, Kabul, Herat, Baki, Ashgabat, Mary, Toshkent, Nukus, Dushanbe, Bishkek, Almaty, Qaraghandy, Astana

b. *Economic regions* (identify with circled letter):

A - African Horn
B - West Bank
C - Mesopotamia
D - Lower Egypt
E - Gaza Strip
F - Vakhan Corridor
G - The Tell
H - Persian Gulf oilfield
I - Jazirah
J - Tengiz Basin
K - Russian Transition Zone

PRACTICE EXAMINATION

Short-Answer Questions

Multiple-Choice

1. Which of the following countries is part of the Maghreb region?

    a) Libya   b) Egypt   c) Algeria
    d) Turkmenistan   e) Turkey

2. Which of the following rivers flows through Iraq?

    a) Euphrates   b) Jordan   c) Blue Nile
    d) White Nile   e) Aral

3. Kemal Atatürk is most closely identified with the city of:

    a) Diyarbakir   b) Mecca   c) Tehran
    d) Ankara   e) Baghdad

4. The Taliban revolution occurred in:

    a) Kazakhstan   b) Afghanistan   c) Egypt
    d) Cyprus   e) Azerbaijan

5. The Ottoman Empire was centered in the modern-day country of:

    a) Palestine   b) Egypt   c) Turkey
    d) Greece   e) Russia

6. Which of the following countries is located in the African Horn?

    a) Syria   b) Senegal   c) Yemen
    d) Sudan   e) Somalia

True-False

1. The West Bank region is hotly contested by Israel and Syria today.

2. The region known as the African Transition Zone contains sizable desert areas.

3. Hierarchical diffusion involves the spreading of an innovation across a large area, "trickling-down" within a country's overall urban system.

4. Both Iran and Iraq are dominated by Shi'ite Muslims.

5. Kazakhstan's capital is named Astana.

6. The Persian Gulf and Mediterranean Sea are connected by the Suez Canal.

Fill-Ins

1. _____ diffusion depends upon a migrating population to transmit an innovation.

2. The _____ Muslim sect dominates life in Iran today.

3. The West Bank is so named with respect to the _____ River.

4. The capital of Egypt is _____.

5. The mountains located between the Black and Caspian Seas are named the _____.

Matching Question on the Realm's Countries

____1. Home of the *fellaheen*   A. Israel
____2. Maghreb kingdom   B. Syria
____3. Capital is Tripoli   C. Uzbekistan
____4. Headquarters of Ottoman Empire   D. Turkey
____5. Adis Abeba   E. Libya
____6. Garagum Canal   F. Saudi Arabia
____7. Jubail planned city   G. Kazakhstan
____8. Shi'ite sect dominant   H. Algeria
____9. Occupies the heart of Turkestan   I. Iraq
____10. Invader of Kuwait in 1990   J. Jordan
____11. Zionist movement   K. Egypt
____12. Tell Atlas   L. Morocco
____13. Tengiz Basin oilfield   M. Turkmenistan
____14. Refugees outnumber natives   N. Iran
____15. Lost Golan Heights to Israel   O. Ethiopia

Essay Questions

1. Geographic definitions of this realm have been difficult to arrive at and are still not entirely satisfactory. Compare and contrast the "Arab World" and "Islamic World" definitions, highlighting the strengths and weaknesses of each. Does the label "Middle East" add anything useful to the argument, and why is it not used in the text as the realm's name?

2. The influence of the Muslim faith has been a dominant shaping force in the history of this realm for the past 1400 years. Elaborate on the diffusion and acceptance of Islam in North Africa/Southwest Asia since AD 600, highlighting cultural-geographical similarities and differences.

3. Discuss the regional problems of the Maghreb, highlighting environmental variables, internal political differences, resource distributions, and economic development opportunities.

4. Discuss the changing political geography of Israel since its creation in 1948. How has this country strengthened itself since the 1967 War, and what is the present likelihood of a more stable coexistence with its Arab neighbors?

5. Compare and contrast the political and economic experiences of Saudi Arabia and Iran in the past ten years, and evaluate the chances for meaningful modernization and development in the two countries during the coming few years.

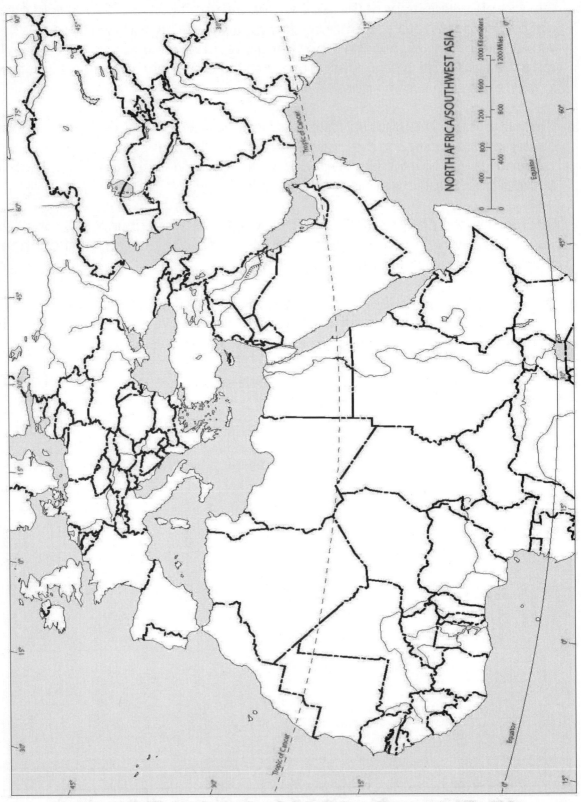

NORTH AFRICA/SOUTHWEST ASIA

Tropic of Cancer

Equator

2000 Kilometers
1200 Miles
0   400   800   1200   1600
0   400   800

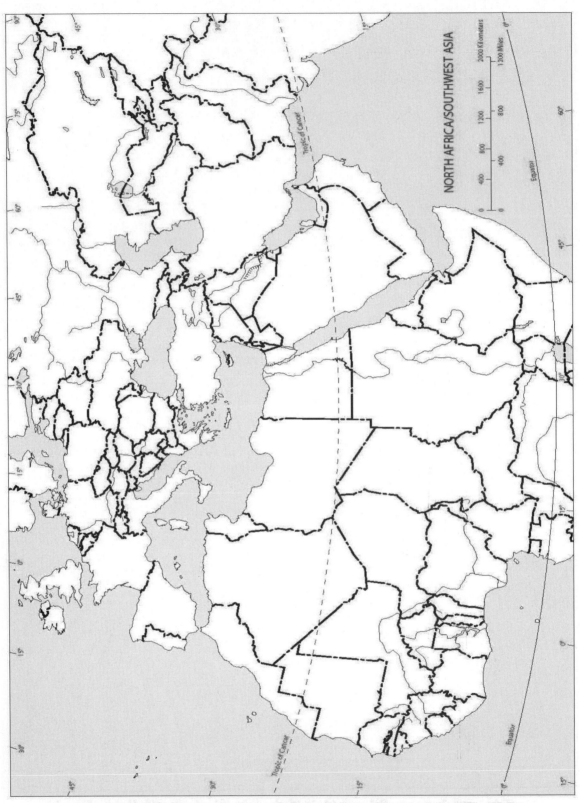

NORTH AFRICA/SOUTHWEST ASIA

Tropic of Cancer

Equator

2000 Kilometers
1200 Miles

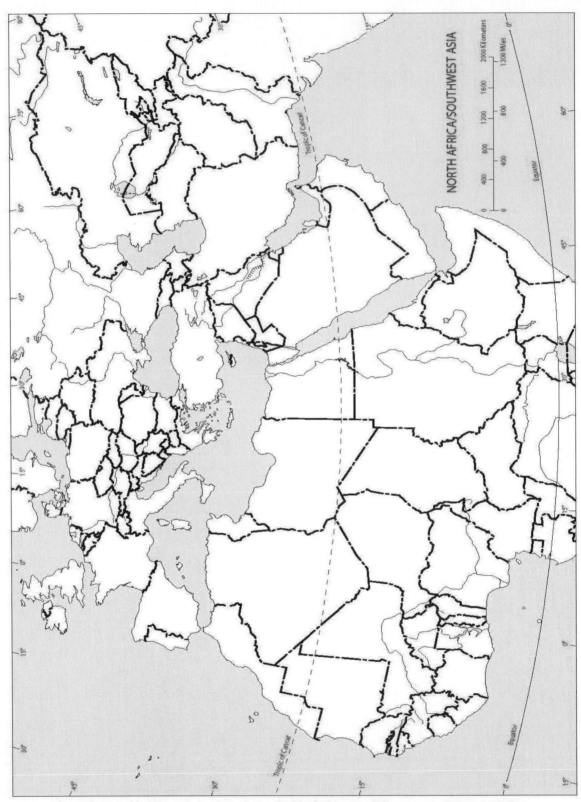

NORTH AFRICA/SOUTHWEST ASIA

400 800 1200 1600 2000 Kilometers

400 800 1200 Miles

Tropic of Cancer

Equator

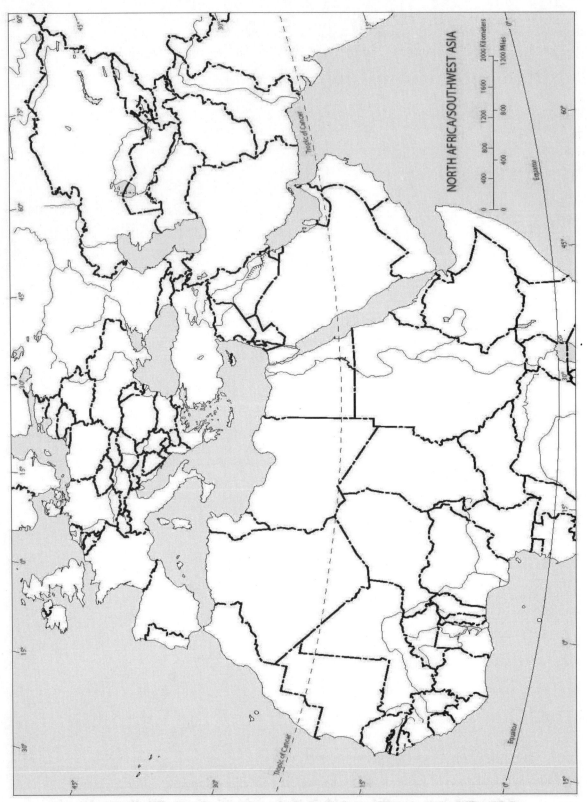

NORTH AFRICA/SOUTHWEST ASIA

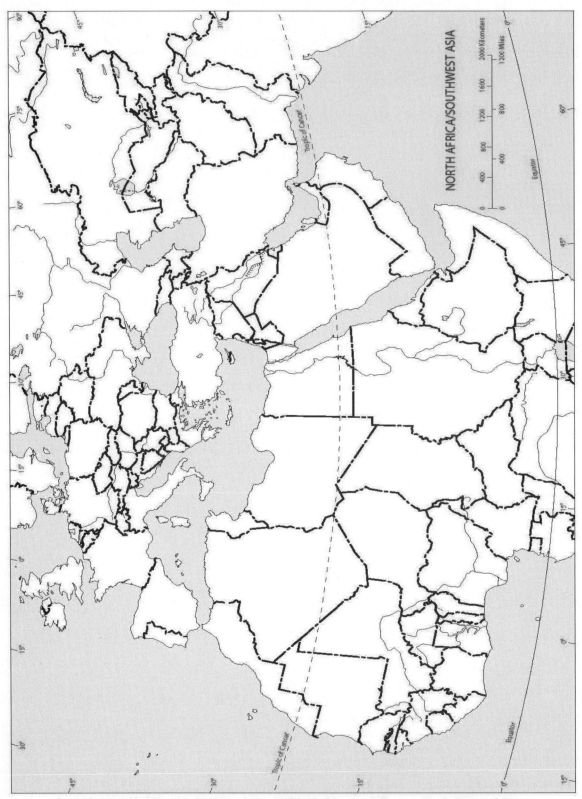

NORTH AFRICA/SOUTHWEST ASIA

Tropic of Cancer

Equator

2000 Kilometers

1200 Miles

CHAPTERS 8 A&B
SOUTH ASIA

OBJECTIVES OF THESE CHAPTERS

South Asia is now the world's largest population agglomeration, with nearly 1.6 billion people residing in 2011 in India, Bangladesh, Pakistan, Sri Lanka, Nepal, Bhutan, and the Maldives. Following the introduction, the rich historical geography of South Asia is traced, highlighting its significant role as a hearth of civilization and its transforming colonial experience under the British. The regional survey of South Asia begins with Pakistan and the multiplying challenges it faces. India, the centerpiece of the realm, is next, and its political, cultural, and economic geographies are reviewed. Then come surveys of Bangladesh, Nepal and Bhutan, Sri Lanka, and the Maldives, treating geographic dimensions as well as current problems faced by each country.

Having learned the regional geography of South Asia, you should be able to:

1. Describe the physiography and the monsoon dynamics of the Indian subcontinent.

2. Trace the major stages of the realm's historical geography, particularly the dramatic transition from colonialism to independence.

3. Understand the opportunities for progress that exist in Pakistan, but are tempered by explosive internal and external political problems.

4. Understand the unique federal political structure of India and its underlying cultural mosaic.

5. Appreciate the crisis of the realm's population numbers (particularly India's).

6. Trace the geography of India's accelerating economic development as it has affected agriculture, manufacturing, and urbanization.

7. Understand the particular miseries of life throughout Bangladesh, and the limited prospects this country faces in the foreseeable future.

8. Understand the forces that buffet Sri Lanka as it tries to emerge from the long conflict between its majority Sinhalese and the minority Tamils.

9. Locate the major physical, cultural, and economic-spatial features of the realm on an outline map.

Monsoon (405-406)

The rainy season produced by an onshore airflow that dominates for weeks, occurring in the hot summer months when low atmospheric pressure over the land sucks in moisture-laden air from the adjacent cooler ocean; especially pronounced in India and Bangladesh.

Social stratification (408)

In a layered or stratified society, the population is divided into a hierarchy of social classes. In an industrialized society, the working class is at the lower end; elites that possess capital and control the means of production are at the upper level. In the traditional caste system of Hindu India, the "untouchables" (also known as *dalits)* form the lowest class or caste, whereas the still-wealthy remnants of the princely class are at the top.

Caste system (408)

The strict segregation of people according to social class in Hindu society, largely on the basis of ancestry and occupational status.

Indo-European (409)

The major world language family that dominates the European **geographic realm** (Fig. 1A-7). This language family is also the most widely dispersed globally (Fig. G-9), and about half of humankind speaks one of its languages.

Dravidian (409)

The language family, indigenous to the South Asian realm, that dominates southern India today; as opposed to the Indo-European languages, whose tongues dominate northern India.

Refugees (412)

People who have been dislocated involuntarily from their original place of settlement.

Forward capital (414)

Capital city positioned in actually or potentially contested territory, usually near an international border; it confirms the **state's** determination to maintain its presence in the region in contention.

Terms of trade (417)

In international economics refers to an agreement between trading partners that stipulates the quantity of imports that can be purchased by one country through the sale of a fixed quantity of exports to the other.

Partition (412)

The division of British India, upon independence in 1947, into India and the two (i.e., West and East) Pakistans along religious lines—with Hindus dominant in the former and Muslims in the latter two.

Refugee (412)

People who have been dislocated involuntarily from their original place of settlement.

Neoliberalism (418)

South Asian economic trend toward privatization, lowered tariffs, reduced government subsidies, corporate tax cuts, and overall deregulation.

Population geography (421)

The field of geography that focuses on the spatial aspects of demography and the influences of demographic change on particular places.

Population distribution (421-422)

The way people, collectively and individually, have spatially arranged themselves in their overall environment.

Population density (421-422)

The frequency of occurrence of a phenomenon within a given area.

Arithmetic density (422)

The number of people per unit area.

Physiologic density (422; 448)

The number of people per unit area of cultivable or arable land.

Fertility rates (422)

The number of births per woman.

Demographic transition (422)

Four-stage model of the changes in population growth exhibited by countries undergoing modern industrialization. High birth and death rates (Stage 1) are followed by plunging death rates (Stage 2) and then a delayed decline in birth rates (Stage 3)— producing a huge net population gain. This is followed by the convergence of birth and death rates (Stage 4) at a low overall level. See Fig. 8A-10on p. 422.

Demographic burden (423)

The proportion of the population that is too young or old to be productive and must be cared for by the productive population.

Population pyramids (423)

Graphic representation or *profile* of a national population according to age and gender. Such a diagram of age-sex structure typically displays the percentage of each age group (commonly in five-year increments) as a horizontal bar, whose length represents its relationship to the total population. Males in each five-year age cohort are shown to the left of the center line; females to the right. See Figure 8A-12.

Sex ratio (424)

A **demographic** indicator showing the ratio of males to females in a given population.

Hindutva (438)

"Hinduness" – expression of Hindu nationalism, heritage, or patriotism.

Chapter 8B

Communal tension (437)

Persistent stress among a country's sociocultural groups that can often erupt into communal violence. India is particularly noteworthy in this regard: frequent conflicts among its highly diverse cultural communities are generally based on (politicized) religion, but they can also be **caste**-based.

Caste system (439)

The strict **social stratification** and segregation of people—specifically in India's Hindu society—on the basis of ancestry and occupation.

Outsourcing (443)

Turning over the partial or complete production of **a** good or service to another party. In economic geography, this usually refers to a company arranging to have its products manufactured in a foreign country where labor and other costs are significantly less than in the home country (which experiences a commensurate loss in jobs)

Informal sector (445)

An unregistered business, which pays no rent or taxes, uses family labor, and often has been passed down through generations.

Double delta (447)

South Asia's combined **delta** formed by the Ganges and Brahmaputra rivers. All of Bangladesh lies on this enormous deltaic plain, which also encompasses surrounding parts of eastern India. Well over 200 million people live here, attracted by the fertility of its soils that are constantly replenished by the **alluvium** transported and deposited by these two of Asia's largest river systems. Natural hazards abound here as well, ranging from the flooding caused by excessive **monsoonal** rains to the intermittent storm surges of powerful cyclones (**hurricanes**) that come from the Bay of Bengal to the south.

Failed state (451)

A country whose institutions have collapsed and in which anarchy prevails. Nepal in the year 2002 is an example.

Buffer state (450)

A country separating ideological or political adversaries. In southern Asia, Afghanistan, Nepal, and Bhutan were parts of a buffer zone between British and Russian-Chinese imperial spheres.

Insurgent state (454)

Territorial embodiment of a successful guerrilla movement. The establishment by anti-government revolutionaries of a territorial base in which they exercise full control; therefore, a state-within-a-state.

Cover the right side of the page with a sheet of paper. Uncover each line after you have attempted to answer the question in the left column. If necessary, refer to textbook page(s) listed at the right.

Question	Answer	Page

Historical Geography

Question	Answer	Page
Who were the Aryans? What was their impact on the emergence of India's modern culture?	Invaders from western Asia who conquered the early Indus Valley civilization beginning around 3500 B.C., but who adopted many of its innovations and pushed settlement frontiers east into the Gangetic Plain and south into the center of the peninsula. India's culture developed from this beginning, including the Hindu religion and its rigid social stratification scheme—the *caste system*.	408
What lasting impact did Buddhism make?	It was dominant during the Mauryan Empire (3rd century B.C.-2nd century A.D.) but soon declined in South Asia, only remaining strong in Sri Lanka (formerly Ceylon) where it still prevails; Buddhism today is mainly centered in East and Southeast Asia (see Intro chapter world religion map).	409
What was the lasting impact of Islam?	After the 10th century it was a strong influence, driving out Buddhism but not Hinduism, which remained dominant in India's Ganges core area; Muslims remain a sizeable minority (slightly less than 15 percent) in India—and form overwhelming majorities in Pakistan and Bangladesh.	410-411
Describe the realm's colonial-era experience.	The British quickly emerged as the dominant colonial power, first through the East India Company and then outright political control from 1857 to 1947. The British introduced many innovations, but forced the colonial economy to become a raw-material producer subservient to the English master.	411-412

Describe the events surrounding the 1947 partition of British India.	The British left India in 1947; however, before withdrawing they separated their former territory into Hindu-dominated India and the two Islamic Pakistans, thereby inducing mass migrations, conflict, and social stresses that still exist.	412

India's Demographic Trends

As of 2011, how large is India's population?	Exceeding 1.2 billion.	434
Is the Demographic Transition Model applicable to India?	Possibly. However, India's population is already much too large, and the pivotal area of the Ganges Lowland remains extremely overpopulated	423

South Asian Characteristics

Which countries constitute this realm?	India, Pakistan, Bangladesh, Nepal, Bhutan, Sri Lanka, and the Maldives. See Fig. 8B-1.	426-map
What is the *lingua franca* of both India and Pakistan?	English.	429

Physical Geography

Describe South Asia's three major physiographic regions.	The northern mountains, the central riverine lowlands, and the southern peninsular plateaus.	430-431
What is the Punjab?	The low hills between the North Indian and Indus plains, the so-called "land of five rivers" astride the India-Pakistan border.	431
What is the course of the Indus River?	It rises in Tibet, crosses Kashmir, bends southward to receive its major tributaries from the Punjab, and then flows southwest through its lower course to reach the Arabian Sea near Karachi.	431

Pakistan

What factors unite Pakistan as a state?	Leaders turned to Islam to bind the country. Pakistan is one of the world's most theocratic states.	433
Where is the country's Muslim stronghold?	In the Pakistani portion of the Punjab, especially around Lahore near the Indian border.	431
What are some examples of irredentism in Pakistan?	The Pushtun refugees along the Afghanistan border in the northwest; the Kashmir dispute further compounds the country's centrifugal forces.	431-432

India's Political Geography

Describe India's federal political structure.	This most populous of the world's federal states has 28 States, 6 Union Territories, and 1 National Capital Territory, with boundaries largely drawn to respect linguistic differences.	434
What is India's present linguistic status?	Hindi is the most widely spoken of the fourteen "official" languages, used by over one-third of the people (there are hundreds of local languages); English has emerged as the *lingua franca*, the chief language of the government and business world. See Fig. 8A-5 on p. 410.	435; 410-map
How do centripetal and centrifugal forces operate in India today?	A persistent centrifugal force is the stratification of society into castes. Centripetal forces include Hinduism, democracy, good communications, and flexibility to accommodate the country's many regional groups (at least through the first decade of the 21st century).	435-440

India's Economic Development

Why is Indian agriculture so stagnant?	Because inefficient traditional farming methods are widespread, transportation is poor, and because land continues to be divided into very small plots.	440

| How has industrialization progressed? | Slowly, despite good coal and iron ore deposits; investments continue, and development in the south around Bangalore has occurred. The information-technology industry is taking off in Bangalore and the suburbs of Delhi and Mumbai. Outsourcing and the growth of software companies and IT services has been enormous. Much of the economic benefit has been confined to a small urban elite sector. | 442-443 |

Bangladesh

| Describe the country's agricultural situation. | Fertile alluvial soils permit highly intensive farming, with rice and wheat for subsistence. Bangladesh remains a nation of subsistence farmers. | 448 |
| What are some problems with Bangladesh faces? | It is highly susceptible to natural hazards and the demographic burden is staggering. | 448-449 |

Nepal and Bhutan

| What problems challenge Nepal? | Centrifugal social and political forces, as well as environmental degradation and overcrowded farmlands. | 450 |
| Bhutan is a buffer state between which two nations? | India and China's Tibet. | 450 |

Sri Lanka

| Which two groups are in conflict here? | The Buddhist, Aryan, *Sinhalese* majority vs. the Hindu, Dravidian, *Tamil* minority. | 451-453 |
| What is the situation in the Jaffna Peninsula? | The Tamil *insurgent state* was overtaken by the Sri Lankan military in 2009. | 454 |

MAP EXERCISES

Map Comparison

1. Compare the physiographic maps of South Asia and Africa, noting key similarities and differences. Describe the Indian subcontinent's position within the continental drift scheme (pp. 285-286) and its present situation within the tectonic-plate framework (Fig. G-4, pp. 13-14).

2. The linguistic map of India is a result of its cultural evolution over the past 3000 years. Analyze that map (Fig. 8A-5, p. 410), accounting for its major differences within the broad framework of the historical-geographic forces that shaped the subcontinent's division into Indo-Aryan and Dravidian cultural spheres.

3. Environmental conditions are frequently associated with crop concentrations in subsistence economies. Compare the map of India's agriculture (p. 442) with the distributions and climates (pp. 16-17), and make observations about the relative productivity of India's regions and the associations of the following crops with specific climate types: cotton, rice, and wheat. Do your conclusions also hold for Sri Lanka (map p. 452)?

4. To get an idea of the 1931-1951 shifts of India's Muslim population and the results of Partition in 1947, compare the two maps of Fig. 8A-6 (p. 413), and describe the overall trends in a spatial context.

Map Construction *(Use outline maps at the end of this chapter)*

1. In order to familiarize yourself with South Asian physical geography, place the following on the first outline map:

    a. *Rivers*: Indus, Ganges, Brahmaputra, Hooghly, Meghna, Godavari, Tapti, Narmada, Mahanadi, Yamuna, Sutlej

    b. *Water bodies*: Bay of Bengal, Arabian Sea, Palk Strait, Ganges-Brahmaputra Delta, Rann of Kutch

    c. *Land bodies*: Sri Lanka, Sind, Kathiawar Peninsula, Cardamom Upland, Thar Desert, Deccan Plateau, Malabar Coast, Coromandel Coast, Konkan Coast, Golconda Coast, North Indian Plain, Punjab, Indus Plain, Baluchistan, Asom Uplands, Chota Nagpur Plateau

    d. *Mountains*: Himalayas, Hindu Kush, Karakoram Range, Western Ghats, Eastern Ghats, Vindhya Range, Khasi Hills

2. On the second map, political-cultural information should be entered as follows:

   a. Draw in the internal State boundaries of India (text p. 434).

   b. Draw in the language regions mapped on p. 410 and color them in appropriate shades.

   c. Compare these two maps and offer conclusions. As an optional additional exercise, consult Fig. 8A-6 (p. 413) and map in concentrations of Muslim minorities within India after Partition; how do these clusters relate to linguistic patterns and how have they changed since 1931 (Fig. 8A-5 [p. 410])?

3. On the third outline map, urban-economic information should be entered as follows:

   a. *Cities* (locate and label with the symbol ●): Mumbai (Bombay), Kolkata (Calcutta), Chennai (Madras), New Delhi-Delhi, Bengaluru, Ahmadabad, Varanasi, Hyderabad (once each for India and Pakistan), Amritsar, Kanpur, Pune, Bhilai, Bhopal, Patna, Jammu, Srinagar, Agra, Chandigarh, Nagpur, Mysore, Pondicherry, Madurai, Jamshedpur, Prayagraj (Allahabad), Lucknow, Jaipur, Karachi, Lahore, Rawalpindi, Islamabad, Multan, Peshawar, Faisal-abad, Dhaka, Cox's Bazar, Chittagong, Khulna, Kathmandu, Thimphu, Colombo, Jaffna, Anuradhapura, Maale

   b. *Economic Regions* (identify with circled letter):

   A - Punjab
   B - Asom
   C - Bihar-Bengal Industrial Region
   D - Chota Nagpur Industrial Region
   E - Maharashtra-Gujarat Industrial Region
   F - Delhi conurbation
   G - Sind
   H - Baluchistan
   I - Hindustan
   J - Vale of Kashmir

PRACTICE EXAMINATION

Short-Answer Questions

Multiple-Choice

1. The majority religion practiced in Sri Lanka is:

   a) Dravidian   b) Aryan   c) Islam
   d) Buddhism   e) Hinduism

2. The partitioning of Hindu India from Muslim Pakistan occurred in:

   a) ca. 460 B.C.   b) 1857   c) 1947
   d) 1971   e) 2006

3. Which of the following is not located in Pakistan?

   a) Malabar Coast   b) Sind   c) Punjab
   d) Baluchistan   e) Islamabad

4. Which Indian city was renamed Chennai?

   a) Calcutta   b) Gandhi   c) Madras
   d) Karachi   e) Bombay

5. Which country does not border India?

   a) China   b) Pakistan   c) Bangladesh
   d) Nepal   e) Afghanistan

True-False

1. The Pushtun refugee irredentist movement in northern Pakistan is based on cultural linkages to neighboring India.

2. Sikhism developed and is still based in the Punjab region.

3. Dravidians are in the majority in the population of the island-nation formerly called Ceylon.

4. The Maldives are located in the Bay of Bengal.

5. Karachi is located at the mouth of the Indus River.

6. The Ganges and Brahmaputra rivers join together to form a double delta in Pakistan.

Fill-Ins

1. The social stratification that dominates Hindu India is known as the _____ system.

2. The narrow lowland of far southeastern India that borders the Indian Ocean and contains the city of Chennai, is called the _____ Coast.

3. The amount of arable land per person is known as the _____ density.

4. The dominant ethnic group of Sri Lanka is called the _____.

5. The body of water to the west of India is the _____ Sea.

6. India's first prime minister after independence was _____.

Matching Question on South Asia

____1. Aryan Buddhist	A.	Sepoy Rebellion
____2. Formerly Calcutta	B.	Chennai
____3. Pakistan's northwestern frontier	C.	Ashoka
____4. Forward capital today	D.	Bangladesh
____5. India's *lingua franca*	E.	Sindh
____6. Dravidian language	F.	Sinhalese
____7. India takeover by	G.	Tamil
British government	H.	Islamabad
____8. Formerly Bombay	I.	English
____9. Indus Delta, centered on Karachi	J.	Kolkata
___10. Mauryan Empire	K.	Mumbai
___11. Disastrous 1991 hurricane	L.	Khyber Pass
___12. Formerly Madras		

Essay Questions

1. The Kashmir problem continues to be an unresolved crisis of major politico-geographical significance to South Asia. Trace the origin of this conflict and its evolution through the early 21st century.

2. Discuss the evolution of Hinduism as the leading religion and social system in India. Trace the various influences and challenges of the Mauryan Empire, the Mogul Empire, the British colonial era, and the course of Hinduism since independence.

3. Discuss the current agricultural geography of India. Draw a sketch map that shows the regional distributions of rice and wheat, and discuss their relationship to the dynamics that shape the country's annual wet monsoon.

4. Bangladesh is undoubtedly plagued by more miseries than almost any other country in the world. Review the environmental, cultural, and economic forces that created this situation, and discuss the relationship between the country's population geography and hopes for future development and modernization.

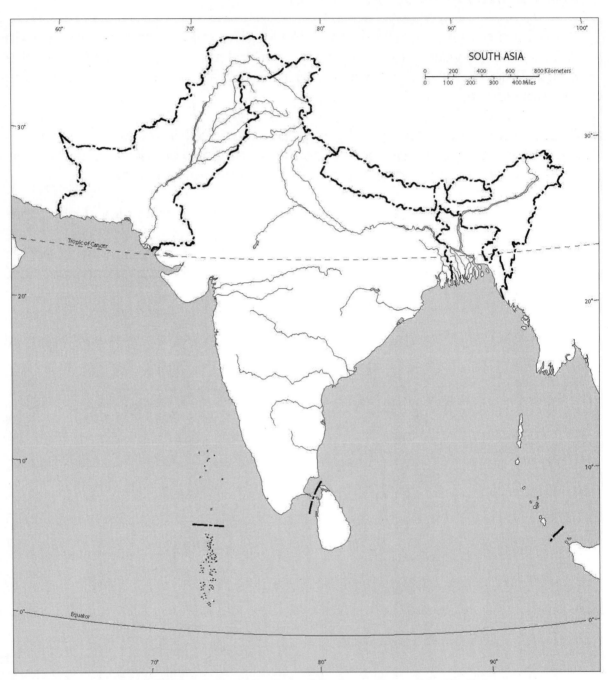

SOUTH ASIA

| 0 | 200 | 400 | 600 | 800 Kilometers |
| 0 | 100 | 200 | 300 | 400 Miles |

Tropic of Cancer

Equator

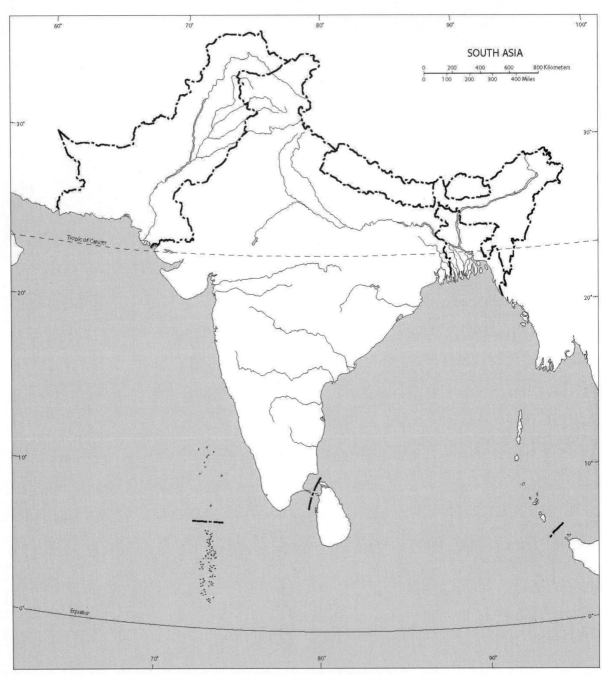

SOUTH ASIA

0   200   400   600   800 Kilometers
0   100   200   300   400 Miles

Tropic of Cancer

Equator

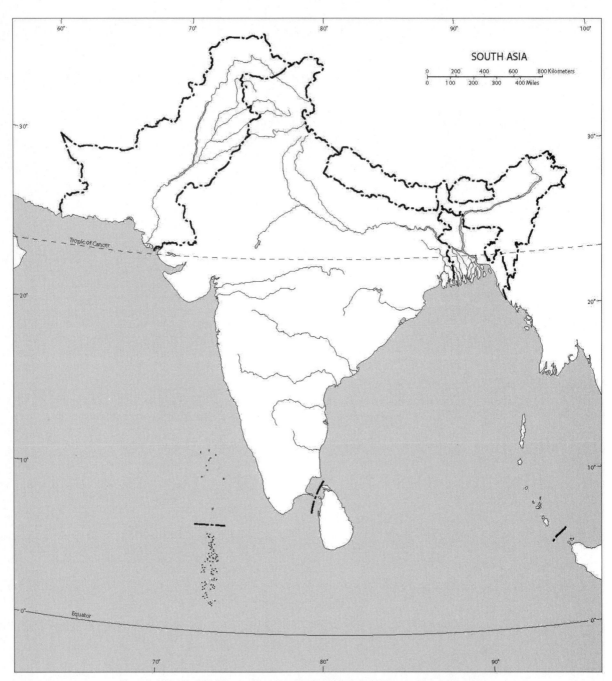

SOUTH ASIA

Tropic of Cancer

Equator

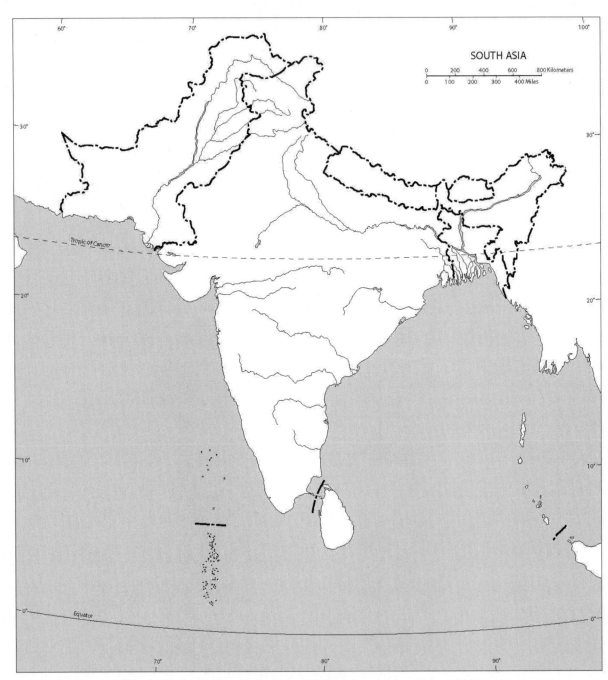

SOUTH ASIA

60° 70° 80° 90° 100°

0   200   400   600   800 Kilometers
0 100 200 300 400 Miles

30° 30°

Tropic of Cancer

20° 20°

10° 10°

0° 0°
Equator

70° 80° 90°

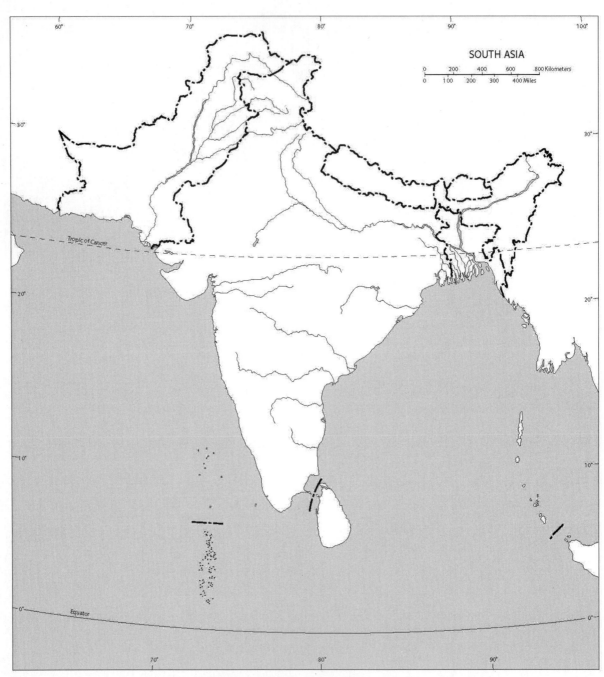

SOUTH ASIA

CHAPTERS 9A&B
EAST ASIA

OBJECTIVES OF THESE CHAPTERS

Chapters 9 A and B cover East Asia, focusing strongly on China—the now-awakened giant that today arouses a great deal of interest and concern. Not only is China the world's largest country in population size and capable of becoming an international superpower in the foreseeable future, it is also in the throes of change. We begin by discussing China's current dimensions, its role as the last surviving empire, and the contradictions that mark its emergence on the global scene. China's natural environment is then discussed. Historical geography follows, tracing the critical events of dynastic China, the convulsions of the colonial era, and the communist transformation. An extended survey of China's regional human geography, highlights the contents of the major subdivisions of the country, with emphasis on the internal structure and functioning of the spatial components of China Proper. We highlight China's developmental experiences since the 1949 Communist Revolution, the modernization program pursued by the post-Mao leadership since the late 1970s, and China's prospects for the immediate future as the leading component of the Asian Pacific Rim. Included is a look at China's booming Special Economic Zones, and the former British Crown Colony of Hong Kong, reunited with China in 1997. We then focus on Japan, North and South Korea, and Taiwan. Japan's rise to a position of global prominence is traced. Working our way along the western Pacific Rim, we next come to South Korea, one of the realm's *economic tigers*. Taiwan, another tiger, follows—its political future remaining to be settled with China (which claims the island-nation).

Having learned the regional geography of East Asia, you should be able to:

1. Understand the basics of the geography of the realm's development and modernization.

2. Appreciate the growing prominence and rapidly changing role of China in global affairs.

3. Appreciate the long tradition of Chinese culture and the tremendous upheavals that its adherents have experienced during the past century and a half.

4. Describe the contents and spatial organization of each of China's major regions, particularly the components of China Proper.

5. Understand the background and goals of China's ongoing modernization —as well as its important spatial expressions—so that you can monitor and evaluate its progress in the future.

6. Understand the Asian Pacific Rim's place in today's world assemblage of regions, composed of discontinuous yet interconnected economic giants and fast-rising economic tigers.

7. Trace Japan's modernization trend and its impact from the Meiji Restoration (1868) to the present.

8. Describe Japan's resource base and the broad regional structuring of its productive activities.

9. Describe the basic geographic contents and spatial organization of South Korea and Taiwan.

10. Locate the leading physical, cultural, and economic-spatial features of the realm on an outline map.

Chapter 9 A

Pacific Rim (458-460)

A far-flung group of countries and parts of countries (extending clockwise on the map from New Zealand to Chile) sharing the following criteria: they face the Pacific Ocean; they evince relatively high levels of economic development, industrialization, and urbanization; their imports and exports mainly move across Pacific waters.

Tsunami (461)

A seismic (earthquake-generated) sea wave that can attain gigantic proportions and cause coastal devastation. The tsunami of December 26, 2004, centered in the Indian Ocean near the Indonesian island of Sumatera (Sumatra), produced the first great natural disaster of the twenty-first century. Our new century's second major tsunami disaster occurred along the coast of Japan's northeastern Honshu Island on March 11, 2011.

State formation (467)

The creation of a **state** based on traditions of human **territoriality** that go back thousands of years.

Dynasty (467)

A succession of Chinese rulers that came from the same line of male descent, sometimes enduring for centuries. Dynastic rule in China lasted for thousands of years, only coming to an end just a bit more than a century ago in 1911.

Neolithic (467)

The "New Stone Age" (ca. 9500-4500 BC), which was the final stage of cultural evolution and technological development among prehistoric humans. Among the many accomplishments that took place, two of the most important for human geography were: (1) the rise of farming through the domestication of plants and animals, and (2) the first permanent village settlements that mark the origin of urbanization.

Confucianism (468)

A philosophy of ethics, education, and public service based on the writings of Confucius (*Kongfuzi*); traditionally regarded as one of the cornerstones of Chinese **culture**.

"People of Han" (470)

Because the Han Dynasty (206 BC-AD 220) was the formative period of China's traditional culture, most ethnic Chinese still refer to themselves in this way.

Sinicization *(Hanification)* (470)

Giving a Chinese cultural imprint; Chinese acculturation.

Extraterritoriality (473)

Legal concept that the property of one state lying within the boundaries of another actually forms an extension of the first. Used in colonial-era China by European diplomats and others to carve out neighborhood enclaves that were off-limits to residents of the host country.

The Long March (473-474)

The year-long exodus of the communists, who were driven out of China Proper in 1934, through the difficult Chinese interior; from their new interior mountain refuge they eventually gathered sufficient strength to emerge again in the late 1940s and finally defeat the Nationalists in 1949.

Manchukuo (475)

The Japanese name for Manchuria during their Empire days of the 1930s and early 1940.

"Great Proletarian Cultural Revolution" (477)

The reign of terror in the name of revolutionary zeal unleashed by Mao Zedong in the middle and late 1960s, wherein the excesses of the militant Red Guards intimidated the population to purge China of noncommunist traditions (and to conceal the disastrous economic setbacks of the "Great Leap Forward").

Asian Tigers (478)

One of the burgeoning beehive countries of the western **Pacific Rim**. Following Japan's route since 1945, these countries have experienced significant modernization, industrialization, and Western-style economic growth since 1980. Three leading economic tigers are South Korea, Taiwan, and Singapore. The term is increasingly used more generally to describe any fast-developing economy.

Special Administrative Region [SAR] (488)

Status accorded the former dependencies of Hong Kong and Macau that were taken over by China respectively, from the United Kingdom in 1997 and Portugal in 1999. Both SARs received guarantees that their existing social and economic systems could continue unchanged for 50 years following their return to China.

Special Economic Zone [SEZ] (491)

Chinese coastal manufacturing and export center, established in the 1980s and 90s, to attract foreign investments and technology transfers through special tax benefits and additional economic incentives.

Economic Geography (494)

The field of geography that focuses on the diverse ways in which people earn a living and on how the goods and services they produce are expressed and organized spatially.

Overseas China (495)

The more than 50 million Chinese who live outside China. Over half live in Southeast Asia and many have become quite successful. A large number maintain links to China, and as investors played a major role in stimulating the growth of SEZs and Open Cities in China's Pacific Rim.

Foreign direct investment (FDI) (495)

A key indicator of the success of an **emerging market** economy, whose growth is accelerated by the infusion of foreign funds to supplement domestic sources of investments.

Economic tiger (502)

One of the burgeoning beehive countries of the western Pacific Rim. Following Japan's route since 1945, these countries have experienced significant modernization, industrialization, and Western-style economic growth since 1980. Three leading economic tigers today are South Korea, Taiwan, and Singapore. Term is increasingly used more generally to describe any fast-developing economy (e.g., Ireland's growing reputation as the "Celtic Tiger").

Buffer state (502)

Remnant of bygone era in which European colonial powers carved out spheres of influence, particularly in Asia. To avoid confrontation, certain territory was left relatively empty between such spheres as a cushion or "buffer." Mongolia evolved as a buffer state between China and Russia. Today it remains landlocked, isolated, and vulnerable, as do other Asian buffer states as Afghanistan and Tibet.

Regional complementarity (504)

Exists when two regions, through an exchange of raw materials and/or finished products, can specifically satisfy each other's demands.

State capitalism (504)

Government-controlled corporations competing under free-market conditions, usually in a tightly regimented society (such as South Korea).

Cover the right side of the page with a sheet of paper. Uncover each line after you have attempted to answer the question in the left column. If necessary, refer to textbook page(s) listed at the right.

Question	Answer	Page
Defining the Realm		
What are the political entities that comprise the East Asian realm?	China, Mongolia, North Korea, South Korea, Japan, and Taiwan.	459-460
The Pacific Rim		
Where have some of the most spectacular developments occurred in the western Pacific Rim?	In Japan, South Korea, Taiwan, coastal and especially southern China.	458
China's Characteristics		
What are China's four great river systems?	Huang He (Yellow), Chang Jiang(Yangtze [Yangzi]), Xi (West), and Liao.	464
What are the main farming / population clusters of China?	The North China Plain, the lower Chang (Yangzi) Basin, the Xi Basin, and the Sichuan (Red)Basin of the interior. See Fig 9A-5.	465-map
Historical Geography		
When and where did the earliest Chinese culture develop?	The Xia Dynasty formed around 2200 B.C., in the core area centered on the lower reaches of the Yi-Luo River Valley.	467
Regions of East Asia		
What are the regions that comprise the East Asian realm?	China Proper, Xizang (Tibet), Xinjiang, Mongolia, and the Jakota Triangle (Japan, South Korea, and Taiwan).	468-470

China Proper

Why was China's self-assured isolationism shattered in the 19th century?	The European powers economically destroyed China's handicraft industries, and their political influence in parts of China bordered on colonialism, a superiority enhanced by encouraging the use of opium that weakened the fabric of Chinese society.	471-472
What is the concept of *extraterritoriality*?	The idea that a foreign power could have territorial enclaves in another country which were legal extensions of that foreign state. Thus foreign states were immune from the jurisdiction of the country in which they were based.	473
List China's political fortunes since 1900.	1900 Boxer Rebellion began the end of colonialist meddling; 1911 overthrow of monarchy led to republican period of government led by Sun Yat-sen and then Chiang Kai-shek; the Japanese dominated from the late 1930s to 1945; civil war from 1945 to 1949 led to communist takeover and Nationalist flight to Taiwan.	473
What was the Long March?	The long, interior exodus of the communists in 1934-1935, who managed to avoid their Nationalist antagonists and eventually emerge victorious in the late 1940s.	473-474
When did communism become the controlling force of the Chinese state?	In 1949, when Mao Zedong proclaimed the success of the revolution and the birth of the People's Republic of China.	476

Political Map and Population

How is China's political framework divided?	There are 22 provinces, 5 autonomous regions, 4 central-government-controlled municipalities (*Shi's*), and 2 special administrative regions.	486
What are China's two largest provinces?	Qinghai and Sichuan.	487

| How large is China's population? | 1.36 billion. | 485 |
| Which ethnic group forms the largest population clusters in China? | The Han Chinese (the ethnic Chinese) form the largest and densest clusters, as shown in Fig. 9A-4, p. 464. | 485 |

China's Regions

What are the leading industrial resources of the Northeast?	Iron ore and coal in the Liao Basin, supporting heavy industrial production in both Anshan and Shenyang; Daqing oilfield near Harbin. See Fig. 9B-9, p. 382.	382-map
Where is the North China Plain, and what is its significance?	The lower basin of the Huang River, extending as far north as Beijing; it is one of the world's most fertile, productive, and densely-populated farming regions.	356
Which of China's major rivers is most navigable?	The Chang Jiang; Shanghai developed at its delta.	356
When and how was Xizang (Tibet) occupied by China?	In 1959, by military force, after years of economic interference and frontier settlement.	377

China's Economic Zones

What are Special Economic Zones (SEZs)?	Manufacturing and export centers in coastal China that lure foreign investment capital and factories through tax benefits and other economic incentives.	491
Besides SEZs, what other economic zones were introduced?	Open Cities and Open Coastal Areas, where foreign investment is encouraged.	491
Why has Shenzhen been the most successful of China's Special Economic Zones (SEZs)?	Shenzhen lies adjacent to Xianggang (Hong Kong), has access to port facilities, and a labor force willing to work for lower wages than in Hong Kong. Japan, the U.S., Taiwan, and Singapore all invest in Shenzhen.	491-493

How did Hong Kong become an economic tiger?	Its large supply of labor and foreign investment helped to launch a textile as well as light manufacturing industries, which exported massively to foreign countries.	488
When did Hong Kong become the Xianggang Special Administrative Region (SAR)?	In mid - 1997, following China's takeover after the departure of the British.	488

Japan's Spatial Organization

Name the four main islands of the Japanese archipelago.	Honshu (the largest), Kyushu, Shikoku, and Hokkaido.	505; 506-map
What are Japan's shortcomings in local industrial resources?	Inferior coal, and very little iron ore; oil supplies are practically nonexistent, forcing importation and heavy reliance on hydroelectric and nuclear energy production.	507
What role did relative location play in Japan's post-World War II development?	The U.S. had become the economic and political focus of the world; Japan's location on the western Pacific coast was no longer remote.	507
What percentage of Japan's land is considered habitable?	An estimated 18 percent.	505
What are the leading features of the Kanto Plain?	Japan's dominant urban and industrial region, focused on Tokyo-Yokohama-Kawasaki; fine harbor; level land and climate suitable for farming; and central location in country.	505
What are the leading features of the Kansai District?	Japan's second-ranking economic region; focused on the Kobe-Osaka-Kyoto triangle; industrial centers, busy ports; good farming area, especially for rice.	505; 507
What are the leading features of the Nobi Plain?	Third-ranking producing region; focused on Nagoya, leading textile manufacturer.	507

What are the leading features of the Kita-kyushu conurbation?	Japan's fourth-largest economic region; focused on 5 cities at northern end of Kyushu Island; favorably situated on trade routes to Korean and Chinese ports.	507

Korea

When and how did Korea become divided into North Korea and South Korea?	In 1945, at the end of World War II, the victorious Allied powers gave Korea north of the 38th parallel to the Soviet Union; territory south of the 38th parallel was given to the United States, forming South Korea.	502
Describe South Korea's growth as an economic tiger.	U.S. and then Japanese aid after the Korean War (1950-1953) launched a development boom that by the 1980s propelled the country to industrial prowess.	504

Taiwan

What types of goods currently dominate Taiwan's exports?	High-technology equipment, such as personal computers, telecommunications components, and precision electronic instruments.	510
What is Taiwan's greatest problem?	Its political situation. In 1949, the Chinese Nationalists made Taiwan the Republic of China. The leaders of the People's Republic of China consider Taiwan to be part of the mainland's territory, its present status that of a wayward province. Taiwan has evolved into a democratic state —its uncertain future hangs in the balance.	510-511

MAP EXERCISES

Map Comparison

1. Discuss the various centripetal and centrifugal forces that currently affect the Chinese state, basing your observations on various maps in Chapter 9— especially the map of ethnic minorities in Fig. 9A-7, p. 469.

2. Compare the map of Japanese manufacturing regions and livelihoods (Fig. 9B-12, p. 506), with its population distribution. Record your observations as to the correspondence between population distribution and the location of the country's primary, secondary, and tertiary industrial regions.

3. Compare the maps of South Korea (Fig. 9B-11, p. 503) and Taiwan (Fig. 9B-13, p. 510). Record your observations as to the internal economic geographies of these economic tigers, noting similarities and differences between spatial patterns of agricultural land use, transportation, and urbanization.

Map Construction (Use outline maps at the end of this chapter)

1. In order to familiarize yourself with East Asia's physical geography, place the following on the first outline map:

    a. *Rivers*: Chang Jiang (Yangzi), Huang He, Wei, Xi, Pearl, Mekong, Brahmaputra, Tarim, Liao, Songhua, Ussuri, Amur

    b. *Water bodies*: East Sea (Sea of Japan), Seto Inland Sea, Tokyo Bay, Tsugaru Strait, Strait of Shimonoseki, Korea Strait, Yellow Sea, Taiwan Strait, Pearl River Estuary, Liaodong Gulf, Korea Bay, Bo Hai Gulf, East China Sea, Formosa Strait, South China Sea, Gulf of Tonkin, Bay of Bengal

    c. *Land bodies*: Hainan Island, Korean Peninsula, Shandong Peninsula, Liaodong Peninsula, Ryukyu Islands, Kurile Islands, Tarim Basin, Gobi Desert, Ordos Desert, Loess Plateau, Qinghai-Xizang Plateau, Takla Makan Desert, Junggar Basin, Qaidam Basin, Red Basin (Chengdu Plain), Liao-Songhua Lowland, Yunnan-Guizhou Plateau, North China Plain, Honshu, Hokkaido, Shikoku, Kyushu, Taiwan

    d. *Mountains*: Pamir Knot, Hindu Kush, Karakoram Range, Himalayas, Trans-Himalayas, Tian Shan, Kunlun, Altun, Altay Mountains

2. On the second map, political-cultural information should be entered as follows:

    a. Draw in all international borders on the map and label each country

    b. Draw in each of China's province-level boundaries and label each province and autonomous region (use Fig. 9B-3 as base)

c.   Plot the concentration of China's ethnic minorities (Fig. 9A-7) and learn their provincial affiliations

3.   On the third outline map, urban-economic information should be entered as follows:

a.   *Cities* (locate and label with the symbol ●): Beijing, Shanghai, Xianggang (Hong Kong), Tianjin, Tangshan, Xian, Wuhan, Guangzhou, Dalian (Lüda), Shenzhen, Shenyang, Changchun, Fushun, Anshan, Harbin, Vladivostok, Lanzhou, Baotou, Taiyuan, Qingdao, Zhengzhou, Nanjing, Hangzhou, Chengdu, Chongqing, Fuzhou, Macau, Haikou, Nanning, Lhasa, Xuzhou, Kunming, Ürümqi, Yumen, Karamay, Yichang, Shantou, Zhuhai, Xiamen, Yantai, Qinhuangdao, Lianyungang, Nantong, Wenzhou, Zhanjiang, Beihai, Ningbo, Tokyo, Yokohama, Osaka, Kobe, Kyoto, Nagoya, Kitakyushu, Sapporo, Seoul, Busan, Kwangju, Pyongyang, Taipei, Kaohsiung

b.   *Economic Regions* (identify with circled letter):

China:
A -   Red Basin (Chengdu Plain)
B -   North China Plain
C -   Liao-Songhua Lowland
D -   Loess Plateau
E -   Pearl River Estuary
F -   Shenzhen Special Economic Zone

Japan:
G -   Kanto Plain
H -   Kansai District
I -   Nobi Plain
J -   Kitakyushu

PRACTICE EXAMINATION

Short-Answer Questions

Multiple-Choice

1. Which movement was launched by Mao Zedong in the 1960s to rekindle enthusiasm for the Chinese brand of communism?

   a) Four Modernizations Program b) Great Proletarian Cultural Revolution
   c) Confucianism (Kongzi)   d) Great Leap Forward
   e) The Long March

2. Which of the following regions is often called "Manchuria" by uninformed foreigners?

   a) Northeast China   b) Taiwan   c) North China Plain
   d) Red Basin of Sichuan e) Xizang

3. Which of the following is not located on the island of Honshu?

   a) Yokohama   b) Kitakyushu   c) Nobi Plain
   d) Kyoto   e) Nagoya

4. Which of the following is not found in the Chang Jiang/Yangzi Basin?

   a) Pudong   b) Xinjiang   c) Chongqing
   d) Wuhan   e) Three Gorges Dam

5. Which of the following is not located in the area of the Pearl Estuary?

   a) Pudong   b) Hong Kong   c) Macau
   d) Xianggang   e) Shenzhen

True-False

1. China is expected to surpass the population total of 2 billion by the year 2015.

2. Despite industrialization, Japan still has a large agricultural sector, employing more than one-third of the labor force.

3. Both Hong Kong and Taiwan have land borders with China.

4. Since 1997 both Hong Kong and Macau have been returned to China's political control.

5. Mongolia is a buffer state lying between China and Japan.

6. The communists' Long March terminated in the Huang (Yellow) Basin.

Fill-Ins

1. The leader of the legendary Long March through interior China was _____.

2. Japan's dominant economic region, which contains its largest city, is the _____ Plain.

3. The economic tiger that has been under the political control of China since 1997 is _____.

4. One of the two Chinese cities designated as a SAR is _____.

5. The fastest-growing city in world history, which now plays a major economic role in the Asian Pacific Rim, is _____.

6. The legal concept whereby the property of one state lying within the boundaries of another forms an extension of the first, is known as _____.

Matching Question on East Asian Places

____1. The Forbidden City
____2. Nationalist Chinese capital
____3. Dalai Lama's former domain
____4. Bordering buffer state
____5. Chinese rustbelt city
____6. Special Economic Zone
____7. South Korea's capital
____8. National belief system
____9. Tokyo's former name
____10. Capital of Taiwan
____11. Binhai New Area
____12. Home of the Manchus
____13. Japanese island
____14. Junggar Basin
____15. Chang Delta

A. Seoul
B. Nanjing
C. Edo
D. Tianjin
E. Xinjiang
F. Shanghai
G. Xizang
H. Northeast China
I. Shenyang
J. Shenzhen
K. Taipei
L. Beijing
M. Mongolia
N. Shinto
O. Hokkaido

Essay Questions

1. Discuss the key developments in the evolution of the Chinese state over the past 4000 years. Emphasize China's rich cultural heritage, explaining why Confucian (Kongfuzi) traditions still persist in many quarters despite the dominance of communism since 1949.

2. The North China Plain is one of the world's most densely settled areas. Discuss the environmental hazards of this subregion, its historical economic geography, and the impact that communism has had on the reorganization of its agricultural production.

3. Compare and contrast two of China's Special Economic Zones. Discuss their purpose and show how each contributes to the Pacific Rim developments that are currently affecting much of China's seaboard.

4. Japan's record of modernization and industrialization over the past century is certainly one of the world's greatest national success stories. Trace the emergence of Japan as an industrial power, emphasizing resources, economic spatial organization, and international relations.

5. Define and discuss the economic tiger concept. Choose one of East Asia's economic tigers of the Pacific Rim, trace its development, and highlight its locational and economic-geographic advantages for sustained national growth.

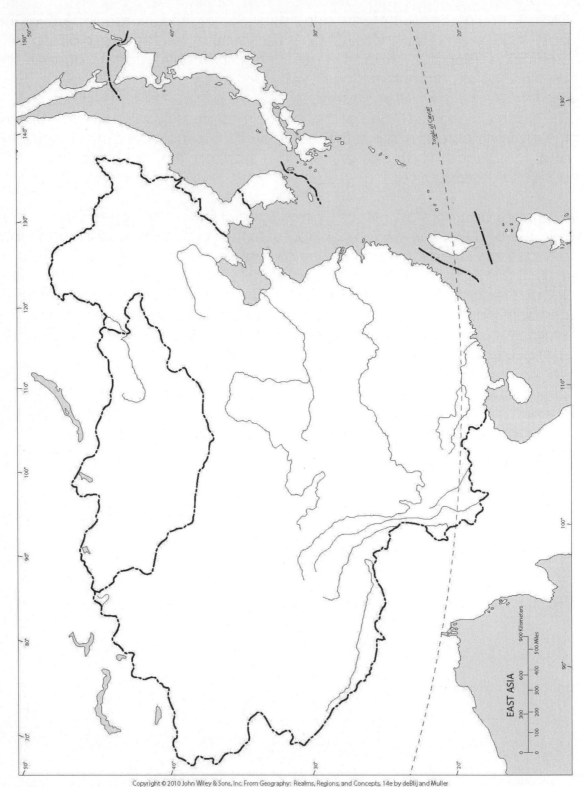

EAST ASIA

900 Kilometers
500 Miles

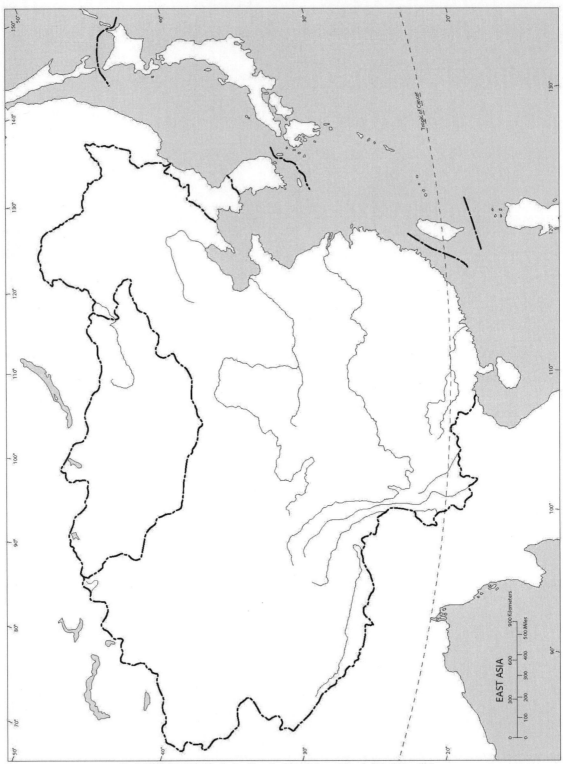

EAST ASIA

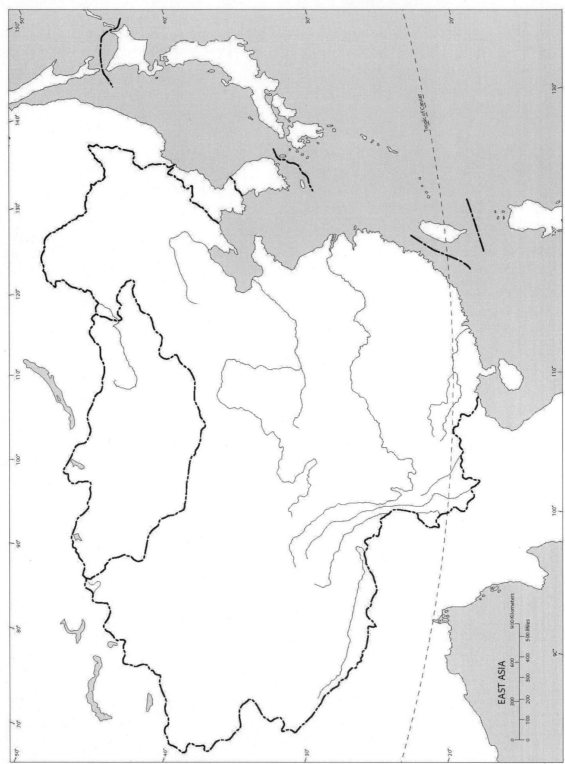

EAST ASIA

900 Kilometers
500 Miles

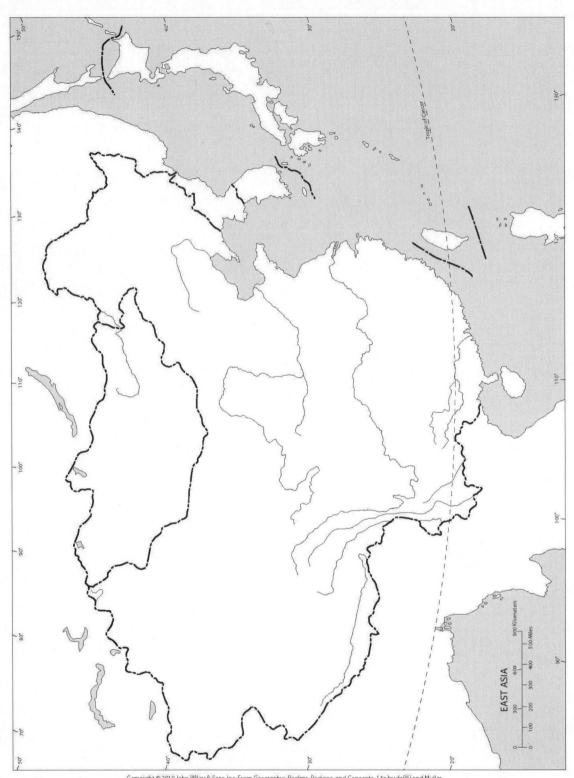

EAST ASIA

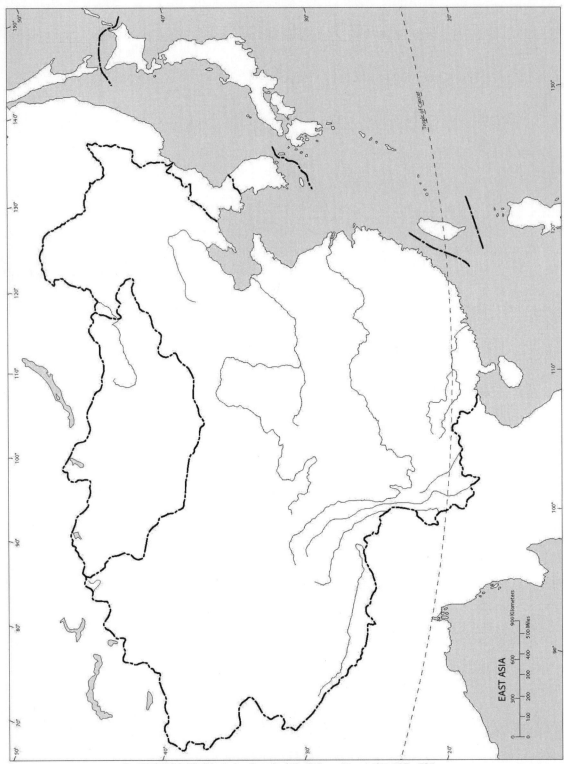

EAST ASIA

CHAPTERS 10 A&B
SOUTHEAST ASIA

OBJECTIVES OF THESE CHAPTERS

Chapters 10 A and B cover Southeast Asia, a realm of such cultural diversity, resulting from centuries of political contests by more powerful neighbors and European as well as Asian outsiders, that it is a classic example of a shatter belt. Following the introduction, population patterns are treated, and the subdivisioning of the realm into mainland and insular components is reviewed. This sets the stage for an overview of European colonial frameworks, highlighting the experiences of the British, the Dutch, and other European countries—including a review of the Chinese experience, which in some ways parallels colonialism. Territorial morphology is discussed at length, and provides the framework for the treatment of the realm's states within the Mainland/Insular rubric.

Having learned the regional geography of Southeast Asia, you should be able to:

1. Understand the basic concepts relating to political boundaries and the shapes of states, as they apply to the realm.

2. Describe the population distributions of the realm's mainland, peninsular, and island components.

3. Appreciate Southeast Asia's cultural-spatial fragmentation and the complex ethnic patterns that have evolved.

4. Grasp the diverse impacts of European colonialism in this part of the world.

5. Locate the major physical, cultural, and economic-spatial features of the realm on an outline map.

Chapter 10A

Buffer zone (516)

A set of countries separating ideological or political adversaries. In southern Asia, Afghanistan, Nepal, and Bhutan were parts of a buffer zone between British and Russian-Chinese imperial spheres. Thailand was a buffer state between British and French colonial domains in mainland Southeast Asia.

Shatter belt (516)

Region caught between stronger, colliding external cultural-political forces, under persistent stress and often fragmented by aggressive rivals. Eastern Europe and Southeast Asia are classic examples.

Wallace's Line (see ch. 11)

As shown in Fig. 11-4, the zoogeographical boundary proposed by Alfred Russell Wallace that separates the marsupial fauna of Australia and New Guinea from nonmarsupial fauna of Indonesia.

Biodiversity (520)

A much higher than usual, world-class geographic concentration of natural plant and/or animal species. Tropical rainforest environments have dominated, but their recent ravaging by **deforestation** has had catastrophic results. The example given in the book is the southern (Costa Rica-Panama) segment of the **land bridge** that forms Central America—over recent geologic time a biogeographical highway for evolutionary change.

Overseas China (524)

There are more than 50 million Chinese who live outside China. Over half live in Southeast Asia (see Fig. 10A-4) and many have become quite successful. A large number maintains links to China, and as investors played a major role in stimulating the growth of SEZs and Open Cities in China's Pacific Rim.

Node (530)

A center that functions as a point of **connectivity** in a regional **network** or **system**. All urban settlements possess this function, and the higher the position of a settlement in its **urban system** or **hierarchy**, the greater its nodality.

Emerging markets (530)

The world's fastest growing national market economies as measured by economic growth rates, attraction of **foreign direct investment**, and other key indicators. Led by the **BRICs** (Brazil, Russia, India, and China), but this club is now expanding to include many other countries.

ASEAN (Assoc. of Southeast Asian Nations) (531)

A center that functions as a point of **connectivity** in a regional **network** or **system**. All urban settlements possess this function, and the higher the position of a settlement in its **urban system** or **hierarchy**, the greater its nodality.

AFTA (ASEAN Free Trade Agreement) (531)

The ASEAN Free Trade Agreement that since 1992, through lowered tariffs and other incentives, has fostered increased trade within Southeast Asia. This is the economic centerpiece of the **Association of Southeast Asian Nations (ASEAN)**, a **supranational** organization whose members include 10 of that realm's 11 states (only East Timor does not participate).

Definition (532)

In boundary-making, a treaty-like document describing in words the location of an international border.

Delimitation (532)

The drawing of a defined boundary on an official map with the approval of the states being divided by that line.

Demarcation (532)

The actual placing of a boundary on the landscape in the form of a fence, cleared strip, or some other physical obstacle.

Physiographic boundary (532)

A boundary that coincides with features of the natural environment, such as rivers or mountains.

Anthropogeographic (ethnocultural) boundary (532)

A boundary that marks the break or transition between cultural groups.

Geometric boundary (532)

Straight-line or curved boundary.

Antecedent boundary (533)

A boundary defined and delimited before the main elements and settlement patterns of the present cultural landscape began to develop.

Subsequent boundary (533)

One that developed contemporaneously with the evolution of the major spatial elements of the cultural landscape.

Superimposed boundary (533-534)

A boundary placed by powerful outsiders on a developed human landscape, usually ignoring pre-existing cultural-spatial patterns.

Relict boundary (534)

One that has ceased to function but whose imprints are still visible on the cultural landscape.

Territorial morphology (534-535)

The size and shape of a state, and what that means in national political life.

Compact state (535)

One possessing roughly circular territory in which the distance from the geometric center to any point on the boundary exhibits little variation; Cambodia is a good example in Southeast Asia.

Protruded state (535)

One possessing territory that is at least in part a narrow, elongated land extension protruding from a more compact core area; in Southeast Asia, the southernmost portions of both Thailand and Myanmar (Burma) are examples.

Elongated state (535)

An attenuated state consisting entirely of territory that is at least six times longer than its average width; in Southeast Asia, Vietnam is a classic example.

Fragmented state (535)

One whose territory consists of several separate, non-contiguous parts, often isolated from one another by international waters or even the land areas of other states; Malaysia, Indonesia, and the Philippines are examples in Southeast Asia.

Perforated state (535)

One that completely encloses another state and is therefore perforated by it; most are small enclaves (such as the Vatican which perforates Italy)—the largest example is Lesotho within South Africa.

Chapter 10B

Domino theory (539)

The belief that political destabilization in one state can result in the collapse of order in a neighboring state, triggering a chain of events that, in turn, can affect a series of contiguous states.

Diffusion (539)

The spatial spreading or dissemination of a **culture** element (such as a technological innovation) or some other phenomenon (e.g., a disease outbreak). For the various channels of outward geographic spread from a source area, see **contagious, expansion, hierarchical**, and **relocation diffusion**.

Protrusion (544)

Territorial shape of a **state** that exhibits a narrow, elongated land extension (or *protrusion*) leading away from the main body of territory. Thailand is a leading example.

Choke Point (550)

A narrow waterway that constrains navigation.

Entrepôt (552)

A place, usually a port city, where goods are imported, stored, and transshipped; a *break-of-bulk* point.

Archipelago (553)

A set of islands grouped closely together, usually elongated into a chain.

Transmigration (558)

Policy of the Indonesian governmental to encourage relocation of population from centrally-located and overcrowded Jawa to the more distant and less populated islands of the country. This aims to reduce population pressure and extend political strength to outlying areas.

Cover the right side of the page with a sheet of paper. Uncover each line after you have attempted to answer the question in the left column. If necessary, refer to the textbook page(s) listed at the right.

Question	Answer	Page

Chapter 10A

Realm Characteristics

Question	Answer	Page
Why is Southeast Asia a shatter belt?	Its cultural-spatial fragmentation results from the collision of stronger outside powers within the realm, much like Eastern Europe.	516
Name the major population concentrations of this realm.	The basins of the Irrawaddy, Chao Phraya, Mekong, and Red Rivers; the fertile volcanic Indonesian island of Jawa; the plantation-studded western coast of the lower Malay Peninsula.	521-522
What percentage of Southeast Asia's population resides in Indonesia and the Philippines?	Just over 50 percent of this realm's population is concentrated in these two countries.	522
Describe the realm's population distribution. Why is it different from the rest of Asia's?	Its leading population clusters are comparatively smaller and lie separated by wide expanses of sparse settlement. Physical obstacles are wide-spread, and agricultural opportunities are much more limited.	520-521

Southeast Asia's Political Geography

Question	Answer	Page
What were the major colonies of the British, French, Dutch, Spanish, and Portuguese?	The British held Burma (now Myanmar), Malaya, and northern Borneo; the French ruled Indochina; the Dutch dominated the East Indies (now Indonesia); as Fig.10A-5 also shows, the Spanish held the Philippines and the Portuguese a part of eastern Indonesia.	527-530; 528-map

Describe the distribution of the Chinese in Southeast Asia.	As Fig. 10A-4 on text p. 523 shows, there is a wide range of penetrations, peaking in western Malaysia, Jawa, the Mekong Delta, western Thailand, Singapore, and the northern portions of Vietnam, Myanmar, and the Philippines.	523-map
What does the term Indochina refer to?	Mainland quadrant of Southeast Asia that French held in colonial times– "Indo" referring to cultural imprints from South & Southeast Asia (Hinduism, Buddhism, Indian art & architecture); "China" signifying the role of Chinese in this realm. Vietnam, Cambodia, & Laos.	527
Describe the origin of the modern state of Malaysia.	Upon independence in 1963, a federation emerged that included Singapore (which severed its ties in 1965); today this state consists of West Malaysia (peninsular Malaysia) and East Malaysia (Sarawak and Sabah on Borneo).	527-529

Political Geography

What is meant by the boundary-related terms of definition, delimitation, and demarcation?	Definition is the document describing the boundary in words; delimitation is its placement upon an official map; demarcation is the actual placement of the border on the landscape.	532
What are the identifying characteristics of antecedent, subsequent, superimposed, and relict boundaries?	Antecedent boundaries precede cultural landscape development; subsequent boundaries emerge together with the elements of the cultural landscape; superimposed boundaries are forced by outsiders with little regard for existing cultural patterns; relict boundaries no longer function, but are still visible on the landscape.	534-535
What is meant by the term *state territorial morphology*?	The size and especially the shape of a country, and what they mean in national political life.	535

What are the identifying characteristics of *compact, fragmented, protruded, elongated,* and *perforated* state territorial morphologies?	Compact states are roughly circular in shape; fragmented states are split into two or more parts, often separated by international waters; protruded states exhibit long, narrow extensions; elongated states are at least six times longer than their average widths; perforated states surround the territory of other states, so that they have a "hole" in them. See Fig. 10A-8, p. 534.	535

Chapter 10B

Regions of the Realm

What countries do the mainland and insular regions of Southeast Asia contain?	Mainland region: Vietnam, Cambodia, Laos, Thailand, Myanmar (Burma); Insular region: Malaysia, Singapore, Indonesia, East Timor, Brunei, and the Philippines	539
List the chief centrifugal politico-geographical forces for the following countries: Vietnam, Cambodia, Myanmar, Malaysia, and Indonesia.	Vietnam was constituted by the separate culture areas of Tonkin, Annam, and Cochin China, and suffered a generation of brutal warfare (from the mid-1940s to the mid-1970s), but is now reunited under communism. Despite Khmer cultural unity, Cambodia has been brutally treated by competing communist factions since 1975; Myanmar is beset by the problems associated with a brutal military government; Malaysia and Indonesia both suffer the effects of politico-spatial fragmentation that include considerable internal cultural variations.	539-559
Why is Singapore an economic tiger?	It has benefited from its relative location on the Strait of Malacca at the tip of the Malay Peninsula; large amounts of crude oil from Southwest Asia are refined; it is a major area of import/export trade; high-tech industries should help to secure a bright future.	551-553
What is the major religion of Indonesia?	Islam—in fact, constituting the world's largest Muslim country; but the faith is taken more casually here, and the islands east of Jawa are dominated by Hindus, Protestants, and Roman Catholics.	557

Where was the Chinese invasion felt most strongly in the Philippines?	In the northern part of the archipelago, especially on the largest island of Luzon.	559
Which religion is dominant in the Philippines?	Roman Catholicism, accounting for 81 percent of the population.	559
What are the three areas of densest population in the Philippines?	(1) The northwestern and south-central part of Luzon; (2) the southeastern extension of Luzon; (3) the islands of the Visayan Sea, between Luzon and Mindanao. See Fig 10A-3 and also Fig. G-8, pp. 20-21.	561

MAP EXERCISES

Map Comparison

1. The importance of the Chinese in this realm has been amply demonstrated in the text. Compare Fig. 10A-5 (p. 526) to the map of Southeast Asia's core areas (Fig. 10A-4, p. 523), and discuss the spatial correspondence between the distribution of ethnic Chinese and the realm's key political decision-making centers.

2. The cultural diversity of Southeast Asia is readily apparent in the map of the realm's ethnic mosaic (Fig. 10A-4, p. 523). Compare and contrast this map to the distribution of religions within Southeast Asia (see Intro chapter world religions map), making observations about the leading regional patterns of ethnic groups and religions.

3. Review the characteristics of antecedent boundaries on text p. 532. Then go back through all of the previous chapters in the text in search of other antecedent boundaries on appropriate maps. Make a list: you should be able to find numerous examples of such international boundaries.

Map Construction *(Use outline maps at the end of this chapter)*

1. In order to familiarize yourself with Southeast Asian physical geography, place the following on the first outline map:

    a. *Rivers*: Irrawaddy, Salween, Mekong, Chao Phraya, Red

    b. *Water bodies*: Indian Ocean, Pacific Ocean, Andaman Sea, Java Sea, Flores Sea, Timor Sea, Banda Sea, Molucca Sea, Celebes Sea, Sulu Sea, Philippine Sea, South China Sea, Gulf of Thailand, Tonkin Gulf, Strait of Malacca, Sunda Strait, Tonle Sap

    c. *Land bodies*: New Guinea, Spratly Islands, Paracel Islands, Hainan Island, Tonkin Plain, Kra Isthmus, Malay Peninsula, Jawa, Sumatera, Borneo, Sulawesi, Bali, Timor, Flores, Sumbawa, Molucca (Maluku) Islands, Mindanao, Luzon, Panay, Samar, Leyte, Palawan, Cebu, Visayan Islands

    d. *Mountains and plateaus*: Annamese Cordillera, Khorat Plateau, Shan Plateau, Arakan Mountains, Yama Range, Dawna Range, Barisan Mountains, Iran Mountains, Muller Mountains

2. On the second map, political-cultural information should be entered as follows:

   a. Label each country and its leading components (as appropriate).

   b. Locate and label each capital city with the symbol *.

   c. Using appropriate color pencils, reproduce the ethnic map (Fig. 10A-4, p. 523), but for the Chinese instead of the color zones on this map use the darker-pink-colored areas shown in Fig.10A-5, p. 526.

3. On the third outline map, economic-urban information should be entered as follows:

   a. *Cities* (locate and label with the symbol ●): Yangon (Rangoon), Mandalay, Naypyidaw, Moulmein, Bangkok, Thon Buri, Chiang Mai, Songkhla, Singapore, Kuala Lumpur, Pinang, Kelang, Alor Setar, Johor Baharu, Kuching, Bandar Seri Begawan, Kota Kinabalu, Banjarmasin, Samarinda, Jakarta, Surabaya, Jogjokarta, Bandung, Padang, Medan, Palembang, Surakarta, Dili, Malang, Pontianak, Ujung Pandang, Manado, Ambon, Phnom Penh, Siem Reap, Viangchan, Louangphrabang, Saigon-Cholon (Ho Chi Minh City), Hanoi, Da Nang, Hué, Loc Ninh, Haiphong, Manila, Cebu, Davao

   b. *Economic regions* (identify with circled letter):

   A -Irrawaddy Delta
   B -Mekong Delta
   C -Tonkin Plain
   D -Chao Phraya Delta
   E -Sarawak
   F -Sabah
   G -Kalimantan
   H -Brunei

PRACTICE EXAMINATION

Short-Answer Questions

Multiple-Choice

1. A boundary delimited before a cultural landscape develops along its route is known as:

    a) antecedent   b) demarcated   c) relict
    d) superimposed   e) subsequent

2. Reunited Vietnam's capital (still mainly called Saigon) is officially named after the communist leader who founded modern North Vietnam, a revolutionary named:

    a) Kota Kinabalu   b) Aung San Suu Kyi   c) Ho Chi Minh
    d) Dien Bien Phu   e) Mao Zedong

3. The largest Muslim country in the world in terms of population size is:

    a) Egypt   b) Indochina   c) Saudi Arabia
    d) Indonesia   e) India

4. Singapore gained independence in 1965 by seceding from:

    a) the Malaysian Federation
    b) the Sunda Archipelago
    c) the Dutch East Indies
    d) the British Commonwealth
    e) the Association of Southeast Asian States

5. Which of the following is not located on the island of Borneo?

    a) Sarawak   b) Jakarta   c) Kalimantan
    d) Brunei   e) Sabah

True-False

1. The city commanding access to the strategic Strait of Malacca is Bangkok.

2. The dominant religion of both Indonesia and Malaysia is Islam.

3. Malaysia is an example of a fragmented state.

4. Singapore's population is dominated by ethnic Chinese.

5. Jakarta is located on the island of Sumatera.

Fill-Ins

1. The ethnic group that constitutes more than three-fourths of Singapore's population is the _____.

2. Both Thailand and Myanmar are examples of states whose territorial morphologies can be classified as _____.

3. The most important island of the Philippines, which contains the capital of Manila, is called _____.

4. The capital of Vietnam is _____.

5. The easternmost island containing Indonesian territory is _____.

Matching Questions on Southeast Asian Countries

____ 1. Most populous Muslim country   A. Brunei
____ 2. Chinese majority   B. Philippines
____ 3. Northeastern neighbor of Thailand C. Cambodia
____ 4. Tagalog speakers   D. Vietnam
____ 5. Islamic sultanate   E. Myanmar
____ 6. Angkor Wat   F. Singapore
____ 7. Chao Phraya core area   G. Indonesia
____ 8. Irrawaddy Delta   H. Laos
____ 9. Contains Mekong Delta   I. Thailand
____10. Sarawak and Sabah   J. Malaysia

Essay Questions

1. The population distribution of Southeast Asia differs markedly from that of the other Asian realms. Compare Southeast Asia's population pattern to India's and China's, discussing the physical, cultural, and economic geographic forces that account for the differences.

2. Review the course of colonialism in Southeast Asia, comparing and contrasting the British, French, and Dutch experiences, and highlighting the politico-geographical patterns that the colonial era has bequeathed to the present.

3. Define the notion of state territorial morphology and discuss its application to Cambodia, Malaysia, and Thailand, respectively, as *compact*, *fragmented*, and *protruded* states.

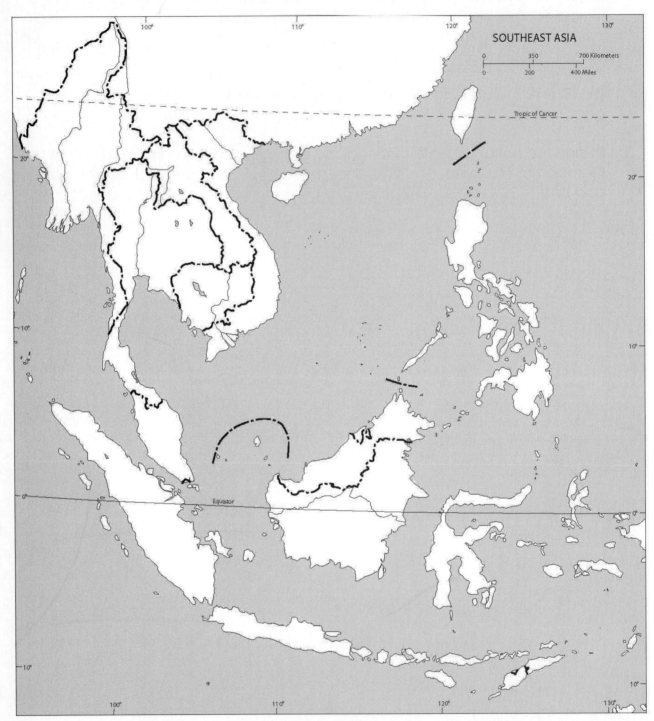

SOUTHEAST ASIA

Tropic of Cancer

Equator

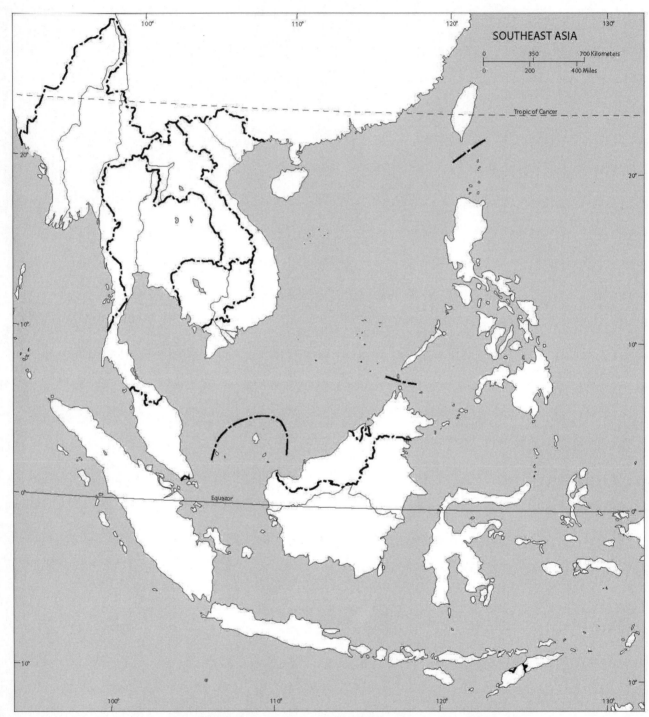

0   350   700 Kilometers

0   200   400 Miles

Tropic of Cancer

100°   110°   120°   130°

20°

20°

10°

10°

Equator

10°

10°

100°   110°   120°   130°

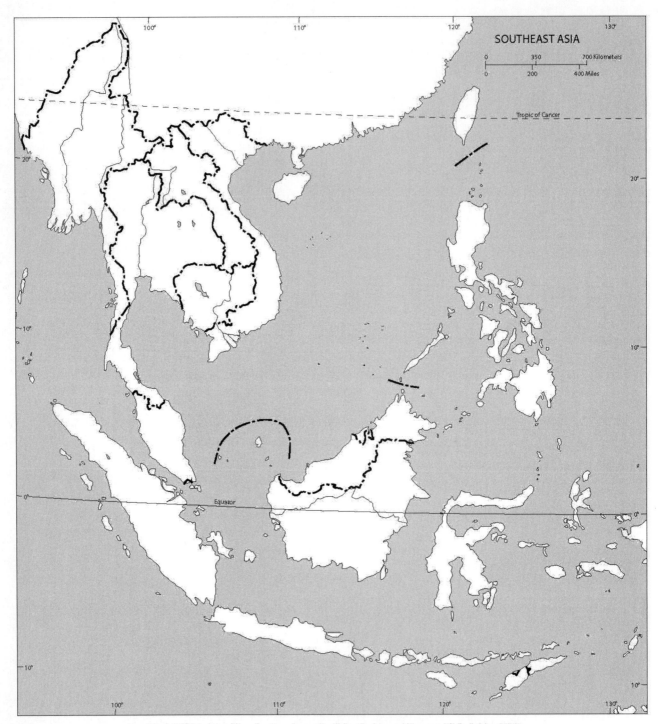

SOUTHEAST ASIA

0   350   700 Kilometers
0   200   400 Miles

Tropic of Cancer

Equator

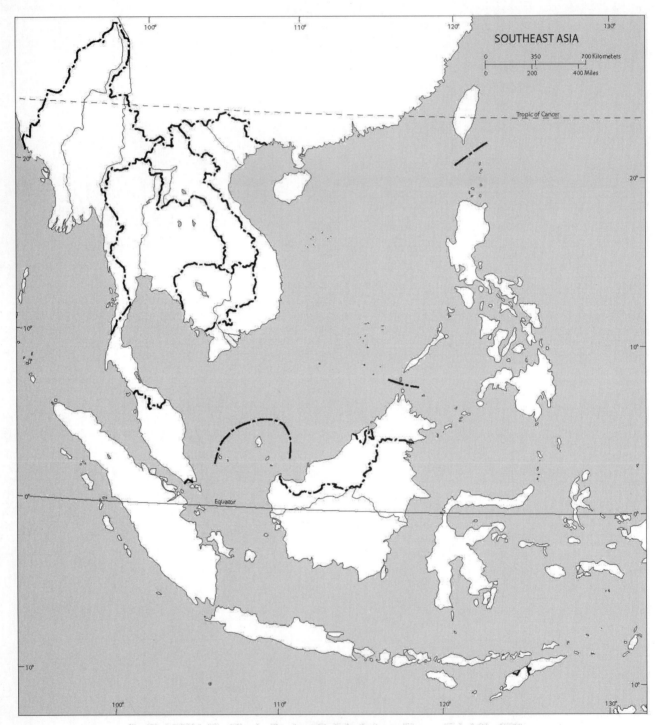

SOUTHEAST ASIA

Tropic of Cancer

Equator

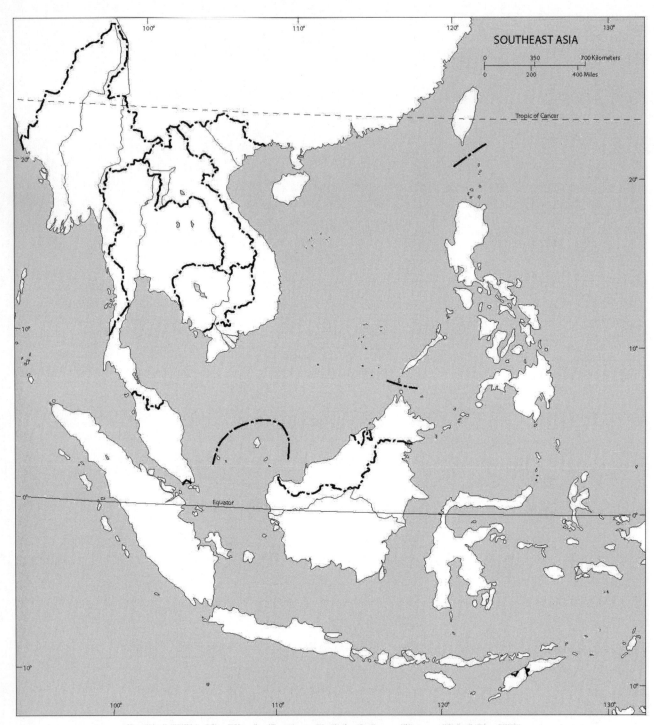

SOUTHEAST ASIA

Tropic of Cancer

Equator

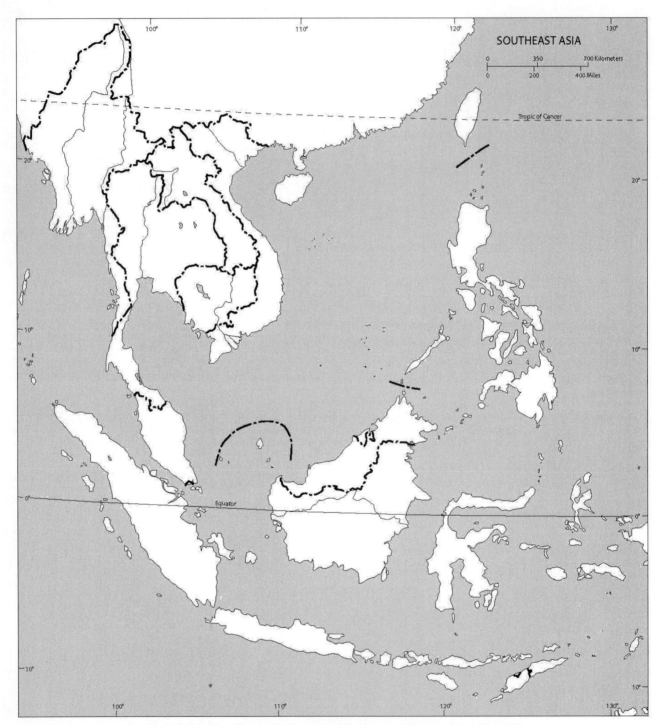

SOUTHEAST ASIA

Tropic of Cancer

Equator

CHAPTER 11
THE AUSTRAL REALM

OBJECTIVES OF THIS CHAPTER

We now move "down under" to survey Australia and its neighbor, New Zealand. An introduction to the realm's natural environment includes a brief description of Australia's unique biogeography. Australia's federal system is then explored in context of the country's modern evolution. The economic geography of Australia is examined next: productive activities, exports, economic problems, and the functional status of the economy within the sphere of the Asian Pacific Rim are some of the issues discussed. Population patterns are then covered, as well as Australia's immigration policies. A discussion of New Zealand's geography ends the chapter, with a description of the country's physical features, economic activities, and ethnic tensions.

Having learned the regional geography of Australia and New Zealand, you should be able to:

1. Understand Australia's political geography and why federalism has succeeded there.

2. Understand the course of Australia's modern evolution, with special emphasis on the country's economic geography.

3. Note the key population patterns and immigration issues of Australia.

4. Understand the main points of New Zealand's human geography.

5. Locate the leading physical, cultural, and economic-spatial features of Australia and New Zealand on a map.

GLOSSARY

Austral (566)

Geographic term meaning "south".

Southern Ocean (568)

The ocean that surrounds Antarctica.

West wind drift (568)

The clockwise movement of water as a current that circles around Antarctica in the **Southern Ocean**.

Biogeography (568)

The study of fauna (animal life) and flora (plant life) in spatial perspective.

Wallace's Line (570)

As shown in Fig. 11-4, the zoogeographical boundary proposed by Alfred Russell Wallace that separates the marsupial fauna of Australia and New Guinea from nonmarsupial fauna of Indonesia.

Aboriginal population (570)

Native peoples, especially in Australia. Often used to designate the inhabitants of areas that were conquered and subsequently colonized by the imperial powers of Europe.

Federation (572)

Derived from the Latin word *foederis*, meaning "association." A political framework wherein a central government represents the various subnational entities (Australia's 6 States and 2 federal territories in this case) within a nation-state where they have common concerns—defense, foreign affairs, and the like—yet permits these entities to retain their own laws, policies, and customs in certain spheres.

Unitary state (572)

Deriving from the Latin word *unitas*, meaning "one." A unified and centralized state, usually with a long tradition of being ruled by a single authority, in which power is exercised equally throughout the country.

Outback (574)

The name given by Australians to the vast, peripheral, sparsely-settled interior of their country.

Import-substitution industries (576)

Industries set up by local entrepreneurs in and near market centers in a remote country (such as Australia) because import costs of foreign goods are extremely high.

Environmental degradation (581)

The accumulated human abuse of a region's natural landscapes that, among other things, can involve air and water pollution, threats to plant and animal ecosystems, misuse of natural resources, and generally upsetting the balance between people and their habitat.

Peripheral development (584)

Spatial pattern in which a country's or region's development (and population) is most heavily concentrated along its outer edges rather than in its interior. Australia, with its peripheral population distribution, is a classic example.

SELF-TESTING QUESTIONS

Cover the right side of the page with a sheet of paper. Uncover each line after you have attempted to answer the question in the left column. If necessary, refer to textbook page(s) listed at the right.

Question	Answer	Page
Land and Environment		
Where does Australia's highest relief occur?	In the Great Dividing Range, mountains that line the eastern coast.	567-568
What area contains much of Australia's mineral wealth?	The Western Plateau and Margins.	568
Biogeography		
What does biogeography study?	It focuses on fauna (animal life) and flora (plant life) in spatial perspective.	568
Who was one of the founders of biogeography as a systematic field?	Alfred Russell Wallace, who did extensive research in Australia.	568-570
What two fields is biogeography divided into?	Phytogeography—focusing on plants; zoogeography–focusing on animals.	568
Australia's Dimensions		
Where is most of Australia's population concentrated?	In the east and southeast, mostly facing the Pacific Ocean. A second, smaller core focuses on the southwest. These regions coincide with the non-desert, humid temperate climate zones.	574
What is meant by the term *Outback*?	The vast, dry, sparsely populated interior of Australia.	574

| When did Australia's indigenous peoples arrive on the island continent? | Aboriginal Australians arrived some 50,000 years ago, developing a patchwork of indigenous cultures. | 571 |

Australia's Political Geography

| When was the Australian Federation formed? | The Commonwealth of Australia was created in 1901. | 572 |
| Why is Australia a federal state? | Because it was formed as a union composed of several separate colonies who wanted to preserve their identities and customs. | 572 |

Australia's Urban and Economic Geography

What percentage of today's Australians reside in cities or towns?	Over 82 percent.	574
How is Australia functionally organized?	Like Japan, large cities are located along the coast, and are the foci for agriculture and manufacturing.	574-575
What are some of Australia's economic problems?	The price of farm products, a major Australian export, fluctuates on the world market; expensive petroleum imports are needed; government-protected industries are backed by strong labor unions; inflation has grown, the national debt has risen, and unemployment has expanded.	576
Name Australia's three leading agricultural exports.	Wool, meat, and wheat.	576
Where are Australia's main dairying zones located?	Close to the large urban markets, as in other areas of the world.	576

What kind of spatial pattern is exhibited by Australia's varied mineral deposits?	Scattered across the country, but in high enough concentrations to make mining profitable. See Fig. 11-6, p. 577.	576-578
What is the present condition of Australian manufacturing?	Concentrated in urban areas; quite diversified; domestically oriented.	578

Australia's Population

Are Australia's European ties as strong as they used to be?	No, a new age is dawning in Australia; its Asian ties are steadily strengthening.	579-581
What percentage of the Australian population is now of British/Irish origins?	Only about 33 percent—down from about 95 percent.	579
Where do most Asian immigrants settle in Australia?	Sydney has the largest contingent; in general, urban areas–immigrants tend to settle in cities and towns.	581

New Zealand

Who are the pre-European people of New Zealand, Polynesians who are still present in large numbers?	The Maori.	582
How do Australia's and New Zealand's landscapes differ?	While Australia's land is of generally low relief and elevation, New Zealand's is generally quite high and has very rugged relief.	582
What is the chief farming region of the South Island?	The Canterbury Plain.	583

MAP EXERCISES

Map Comparison

1. Compare the map of Australian minerals and agricultural areas (Fig.11-6, p. 577) with the country's population distribution (Fig. 11-2, p. 567). What spatial associations are apparent, and for which activities is remoteness a serious problem?

2. Compare the maps of Australia and New Zealand's population/settlement patterns (Fig. 11-2, p. 567 and 11-8, p. 583) to both the map of world climates (Fig. G-7, pp. 16-17) and Australia's physiography (Fig. 11-3, p. 569). What spatial relation-ships can be discerned? Are Australia and New Zealand's patterns of settlement similar or different?

Map Construction (Use outline maps at the end of the chapter)

1. To familiarize yourself with Australia and New Zealand's physical geography, place the following on the first of the outline maps:

   a. *Water bodies*: Murray River, Darling River, Bass Strait, Great Australian Bight, Gulf of Carpentaria, Coral Sea, Tasman Sea, Cook Strait

   b. *Land areas*: Great Dividing Range, Murray Basin, Nullarbor Plain, Great Artesian Basin, Great Sandy Desert, Great Victoria Desert, Arnhem Plateau, Gibson Desert, Tasmania, Southern Alps, Canterbury Plain, North Island, South Island

2. On the second outline map, enter the following politico-geographical information:

   a. Draw in all international borders on the map and label each country.

   b. Label all of the States and Territories of Australia, and label each of their capitals with the symbol □; label the national capital with the symbol *.

3. On the third outline map, enter the following urban-economic information:

   a. *Cities* (locate and label with the symbol ●) : Sydney, Melbourne, Adelaide, Brisbane, Canberra, Hobart, Newcastle, Rockhampton, Alice Springs, Darwin, Perth, Kalgoorlie, Wellington, Auckland, Christchurch, Dunedin

   b. *Economic centers* (identify with a circled letter):

A - Broken Hill	D - Coronation Hill
B -Kambalda	E -Canterbury Plain
C -Mount Isa	

PRACTICE EXAMINATION

Short-Answer Questions

Multiple Choice

1. Australia's governmental structure can best be classified as a:

    a) unitary state   b) federal state   c) dictatorship
    d) monarchy   e) republic

2. The ethnic group that reached Australia approximately 50,000 years ago is known as the:

    a) Aborigines   b) Indians   c) British
    d) Chinese   e) Maori

3. Which city located is located on Australia's western, Indian Ocean coast:

    a) Sydney   b) Auckland   c) Melbourne
    d) Perth   e) Canberra

4. Australia's federal form of government was established in:

    a) 1492   b) 1788   c) 1901
    d) 1927   e) 2001

True-False

1. There was a gold rush in the 1850s in Australia.

2. New Zealand's population, like Australia's, is largely concentrated on its coastal fringes.

3. The Australian Outback is a rugged, highland zone.

4. Australia's percentage of urban population is higher than that of the United States.

Fill-Ins

1. Australia's largest city is _____.

2. New Zealand's larger island is its _____ Island.

3. New Zealand's capital is _____.

4. The capital of Victoria State and Australia's second largest city is _____.

Matching Question on Australia/New Zealand

_____ 1. Australia's only large, non-coastal city	A. Maori
_____ 2. New Zealand's largest lowland	B. Canterbury Plain
_____ 3. Largest minority in New Zealand	C. Auckland
_____ 4. Sugarcane-growing area of Australia	D. Queensland
_____ 5. New Zealand's largest city	E. Canberra
_____ 6. Capital of South Australia	F. Adelaide

Essay Questions

1. Australia is only slowly reorienting itself to become a major player in western Pacific Rim developments. Discuss why this is the case, including such factors as exports, general economic conditions in the country, and its locational situation. Explore these same factors for the leading countries of the Asian Pacific Rim, such as Japan, Singapore, and Taiwan, and make comparisons.

2. There is a growing risk of ethnic polarization in New Zealand. Describe the Maori situation—its historical development, current status, and prospects for the future.

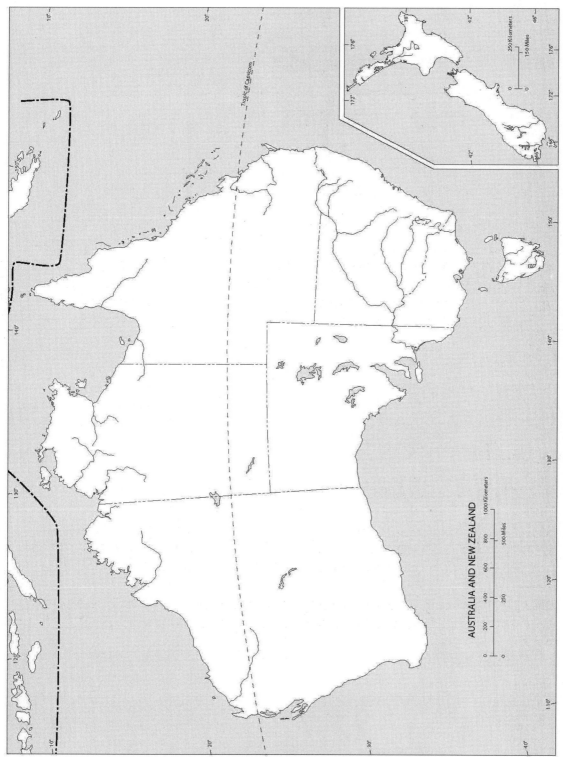

AUSTRALIA AND NEW ZEALAND

Tropic of Capricorn

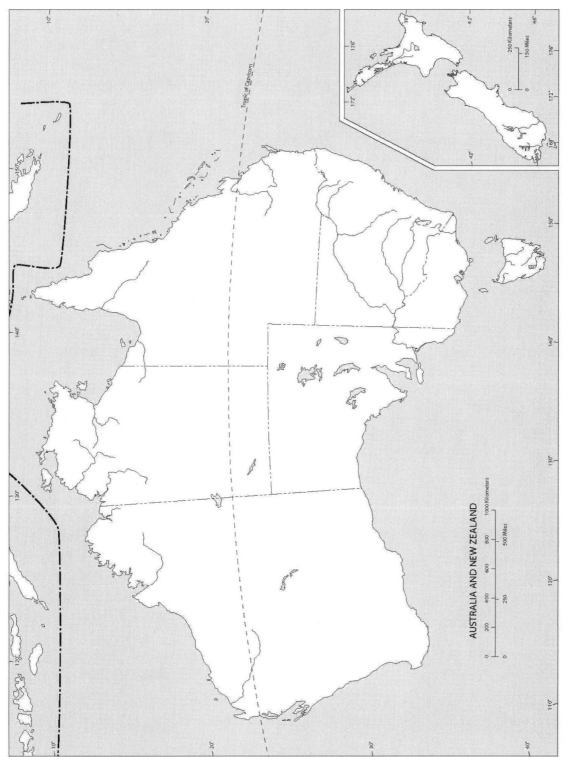

AUSTRALIA AND NEW ZEALAND

Tropic of Capricorn

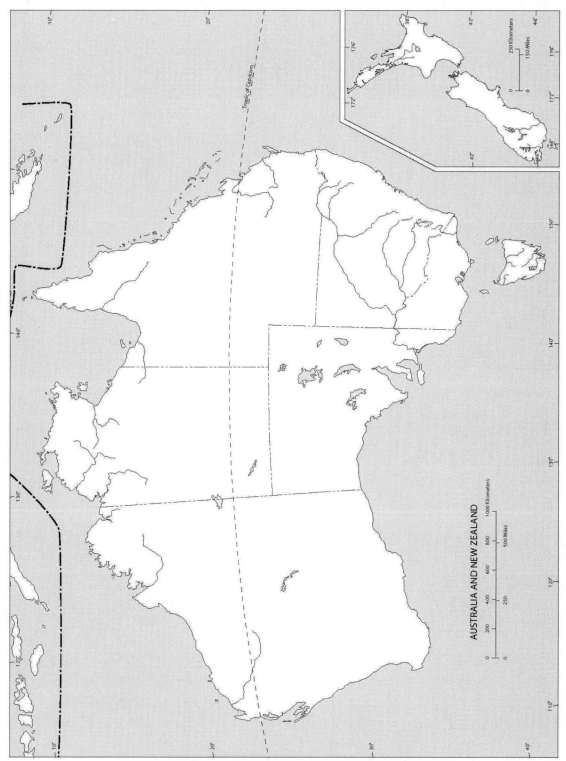

AUSTRALIA AND NEW ZEALAND

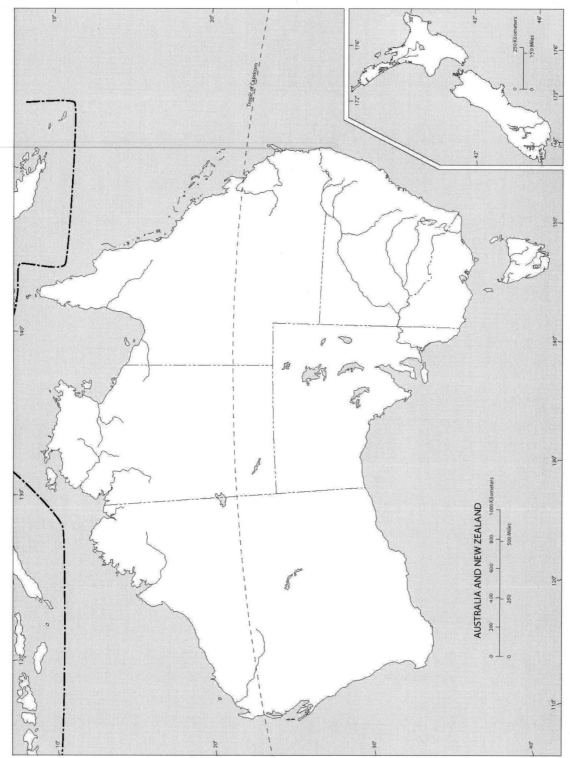

AUSTRALIA AND NEW ZEALAND

Tropic of Capricorn

250 Kilometers

150 Miles

1000 Kilometers

500 Miles

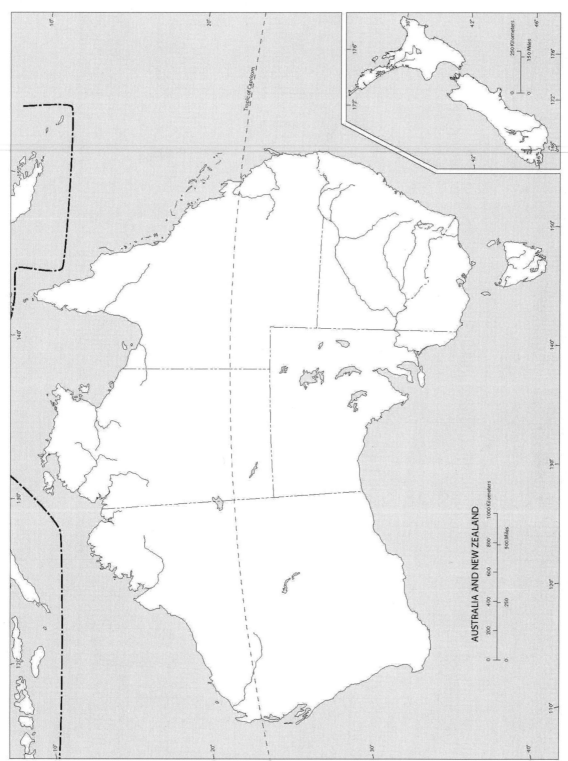

AUSTRALIA AND NEW ZEALAND

Tropic of Capricorn

CHAPTER 12
THE PACIFIC REALM

OBJECTIVES OF THIS CHAPTER

This final chapter covers the largest area by far, greater than any of the previously-treated realms—the vast Pacific Ocean, whose thousands of islands contain inhabitants who dwell in a fragmented, widely dispersed, highly complex geographic realm. Appropriately, the topic of marine geography is introduced, and this is followed by an elaboration of the political geography of maritime claims. We then turn to a survey of the realm's three regions—Melanesia, Micronesia, and Polynesia. The chapter concludes with a brief survey of the Antarctic and Arctic, highlighting environmental and geopolitical issues.

Having learned the regional geography of the Pacific Realm, you should be able to:

1. Comprehend basic concepts of marine geography and maritime claims.

2. Appreciate its fragmented cultural complexities as well as its regional commonalities.

3. Describe the leading geographic characteristics of Melanesia, Micronesia, and Polynesia, and locate their major places on an outline map.

4. Understand this unique realm's opportunities and challenges given its diverse components.

5. Describe the environmental and geopolitical issues that face Antarctica and the Arctic.

GLOSSARY

Marine geography (590)

The geographic study of oceans and seas. Its practitioners investigate both the physical (e.g., coral-reef biogeography, ocean-atmosphere interactions, coastal geomorphology) and human aspects (e.g., maritime boundary-making, fisheries, beachside development) of oceanic environments.

Territorial sea (590)

The water adjoining a country's coastline which is under the legal control of that state.

High seas (590)

Ocean areas free from the legal claim of any state, because they are beyond the defined Exclusive Economic Zones.

Continental shelves (590)

The shallow, sloping seafloor that leads away from a coastline down to about 100 fathoms (600 feet).

Exclusive Economic Zone (EEZ) (591)

An oceanic zone extending up to 200 nautical miles from a shoreline, within which the coastal state can control fishing, mineral exploration, and additional activities by all other countries.

Maritime boundary (591)

An international boundary that lies in the ocean. Like all boundaries, it is a vertical plane, extending from the seafloor to the upper limit of the air space in the atmosphere above the water.

Median line boundary (591)

An international maritime boundary drawn where the width of the sea is less than 400 nautical miles. Because the states on either side of that sea claim Exclusive Economic Zones of 200 nautical miles, it is necessary to reduce those claims to a (median) distance equidistant from each shoreline. Delimitation on the map almost always appears as a set of straight-line segments that reflect the configurations of the coastlines involved.

Melanesia (596-597-map;595-598)

The most populous Pacific region covering New Guinea and the island groups stretching to its southeast (see Fig. 12-3).

Micronesia (596-597-map; 599)

The chains of small islands lying north of Melanesia and east of the Philippines (see Fig. 12-3).

Polynesia (596-597-map;600)

The remainder of the realm, lying within the great triangle circumscribed by New Zealand, Hawai'i, and Easter Island (see Fig. 12-3).

High islands (599)

Volcanic islands of the Pacific Realm which are high enough in elevation to wrest substantial moisture from the tropical ocean air. They tend to be well watered, their volcanic soils enable productive agriculture, and they support larger populations than low islands–which possess none of these advantages and must rely on fishing and the coconut palm for survival.

Low islands (599)

Low-lying coral islands of the Pacific Realm that–unlike high islands–cannot wrest sufficient moisture from the tropical ocean air to avoid chronic drought. Thus productive agriculture is impossible and their modest populations must rely on fishing and the coconut palm for survival.

Antarctic Treaty (602)

Treaty signed in 1961, ensuring scientific research collaboration, environmental protection, and the prevention of military activities in Antarctica.

Ice shelves (602)

Permanent slabs of floating ice in the Antarctic region, such as the Ronne and the Ross. See Fig. 12-4, p. 602.

SELF-TESTING QUESTIONS

Cover the right side of the page with a sheet of paper. Uncover each line after you have attempted to answer the question in the left column. If necessary, refer to textbook page(s) listed at the right.

Question	Answer	Page
Marine Geography		
What politico-spatial problems do marine geographers study?	How far a state's jurisdiction extends off its coast-line; by what method seaward boundaries are defined, delimited, and sometimes demarcated; and what responsibilities states have in the remaining "high seas."	590
What is a state's territorial sea?	The strip of water adjoining a country's coastline, which is legally controlled by that state.	590
What was the outcome of the UNCLOS conferences?	Among key provisions, a 12-mile territorial sea was defined for all countries; a 200-mile Exclusive Economic Zone, in which states have exclusive rights, was also created.	591
Why was it necessary to draw median-line boundaries between many countries' territorial seas?	Many nations are located next to choke points, and thus are separated from each other by less than 24 miles of water. Both 12-mile territorial seas and 200-mile Exclusive Economic Zones would overlap, creating endless maritime boundary disputes.	591-592
Dimensions of the Pacific Realm		
What 3 regions constitute the Pacific Realm?	Melanesia, Micronesia, and Polynesia.	593
When did Papua New Guinea become a sovereign state?	In 1975, after nearly a century of British and Australian administration.	595

What are the two largest cultural groups of Papua New Guinea?	The largest group is the Papuans, who live in the south; the Melanesians are second in terms of population size, and inhabit the north and east of the country.	595
Describe the economic potential of Papua New Guinea.	Oil was discovered in the 1980s, and became the largest export by value in the late 1990s. Gold, silver, and many other materials are now exported, as well as agricultural products, largely to Pacific Rim countries.	596
What does the term *Micronesia* mean?	It is derived from the tiny average size of its islands (*micro* means small).	599
Differentiate between *high-island* and *low-island* cultures.	High-island cultures are associated with volcanic islands whose fertile soils permit farming; low-island cultures are based on dry, low-lying coral islands, and forced to rely on fishing.	599
What does the term *Polynesia* mean?	It is derived from the large number of islands it contains (*poly* means many) within its vast areal extent, as shown in Fig. 12-3, pp. 596-597.	600
Describe the current political geography of Polynesia.	Highly complex. The U.S.'s 50th state comprises the Hawaiian archipelago; Tuvalu, Kiribati, & Tonga are former British dependencies; certain islands remain under French control, including Tahiti; other islands continue to be administered under the flags of Chile, New Zealand, the U.S., & the United Kingdom.	600-601
Do Antarctica and the Southern Ocean constitute a geographic realm?	In physiographic terms, yes, but in functional terms, no. No functional regions, cities, transport networks, etc. have developed.	601
Why are states interested in making territorial claims in remote Antarctica?	Land and sea contain raw materials, such as fuels, minerals, and proteins—which may become crucial reserves in the future.	602
What did the Antarctic Treaty and the Wellington Agreement stipulate?	The Antarctic Treaty was signed in 1961 to protect the environment, keep national claims in abeyance, prohibit military activity, and promote scientific collaboration. The Wellington Agreement, signed in 1991, restricted existing claims to the Antarctic land mass.	602-603

MAP CONSTRUCTION *(Use outline maps at the end of the chapter.)*

1. To familiarize yourself with the Pacific Realm's geography, place the following on the first of the outline maps:

   Islands: New Guinea, Bougainville, Solomon Islands, New Caledonia, Hawaiian Islands (Oahu, Hawai'i, Maui, Molokai, Lanai, Kauai), Guam, Midway Island, Wake Island, Nauru, Fiji, Tonga, Samoa, Tahiti, Easter Island, Pitcairn Island.

   Water bodies: Coral Sea, Arafura Sea, Philippine Sea.

2. On the second outline map, enter the following politico-geographical information:

   Political Units: Papua New Guinea, Solomon Islands, Vanuatu, New Caledonia, Wallis and Futuna, Western Samoa, American Samoa, Tonga, Fiji, French Polynesia, Tokelau Islands, Nauru, Kiribati, Tuvalu, Cook Islands, Easter Island, Niue, Rarotonga, Truk, and—comprising the former U.S. Trust Territory in Micronesia—the Northern Mariana Islands, Republic of Palau, Federated States of Micronesia, and Republic of the Marshall Islands.

3. On the third outline map, enter the following urban information:

   Cities (locate and label with the symbol ▢) : Honolulu, Hilo, Port Moresby, Suva, Pago Pago, Nouméa, Papeete.

PRACTICE EXAMINATION

Short-Answer Questions

Multiple-Choice

1. The major economic activity associated with low-island cultures is:

   a) hunting   b) farming   c) fishing
   d) tourism   e) nomadic herding

2. Which Melanesian island, still ruled by France today, is one of the world's largest sources of nickel?

   a) New Caledonia   b) New Guinea   c) Tahiti
   d) Fiji   e) Bougainville

True-False

1. Papua New Guinea is located in Polynesia.

2. High-island cultures are associated with fertile soils and farming economies.

Fill-Ins

1. The most heavily populated of the three Pacific regions is _____.

2. The largest in areal extent of the three Pacific regions is _____.

Matching Question on the Pacific Realm

_____ 1. Micronesia's largest island   A. Port Moresby
_____ 2. Large-scale copper mining   B. Kiribati
_____ 3. Capital of Papua New Guinea   C. Bougainville
_____ 4. Large majority of Hawaiians located here   D. Oahu
_____ 5. Formerly called the Gilbert Islands   E. Guam

Essay Question

1. Compare and contrast the major geographical dimensions of Melanesia, Micronesia, and Polynesia, and evaluate the potential for future development and modernization in each region.

2. Compare and contrast the major environmental and geopolitical issues that today face the Arctic.

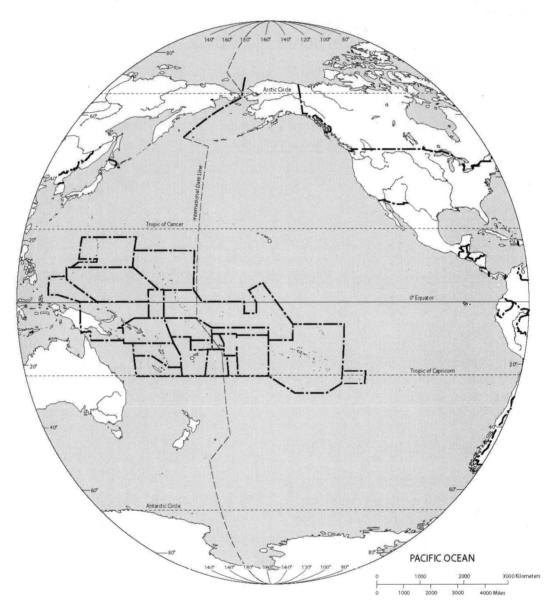

PACIFIC OCEAN

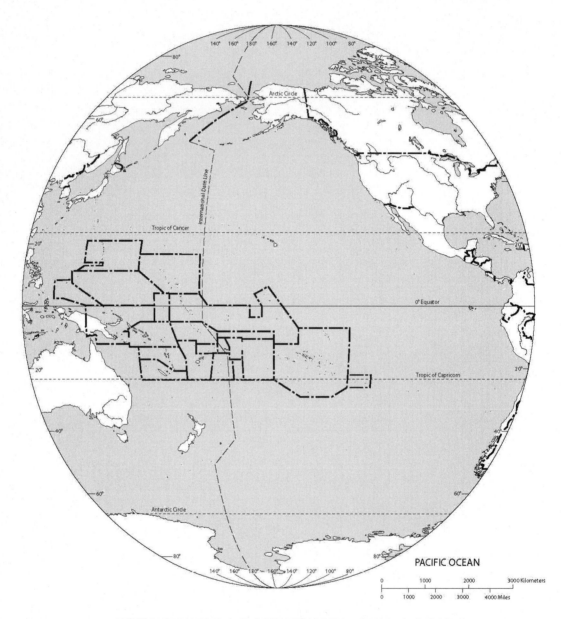

PACIFIC OCEAN

| 0 | | 1000 | | 2000 | | 3000 Kilometers |
| 0 | 1000 | 2000 | 3000 | 4000 Miles |

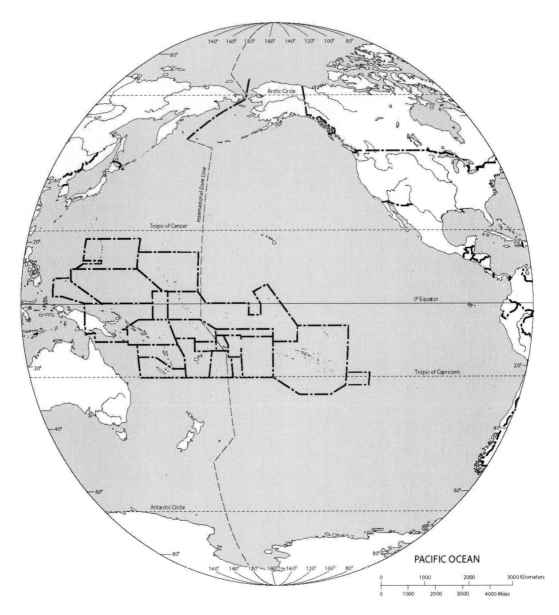

PACIFIC OCEAN

| 0 | 1000 | 2000 | 3000 Kilometers |

| 0 | 1000 | 2000 | 3000 | 4000 Miles |

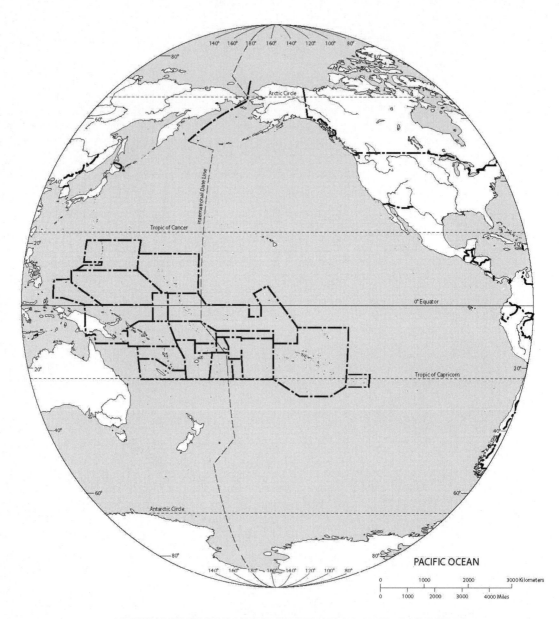

PACIFIC OCEAN

0   1000   2000   3000 Kilometers

0   1000   2000   3000   4000 Miles

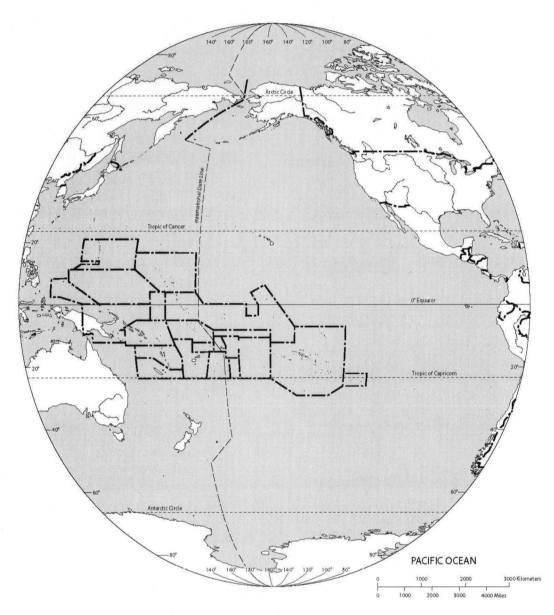

PACIFIC OCEAN

0   1000   2000   3000 Kilometers
0   1000   2000   3000   4000 Miles

ANSWERS TO
PRACTICE SHORT-ANSWER EXAMINATIONS

Introduction

Multiple-Choice:	1(b), 2(a), 3(a), 4(e), 5(c)
True-False:	1(false), 2(true), 3(true), 4(false), 5(true)
Fill-Ins:	1(India), 2(functional), 3(B), 4(cultural landscape), 5(New Zealand), 6(relative)

Chapter 1A&B

Multiple-Choice:	1(d), 2(a), 3(d), 4(d), 5(b) 6(a)
True-False:	1(true), 2(true), 3(true), 4(true), 5(false)
Fill-Ins:	1(Spain), 2(centripetal), 3(devolution), 4(Denmark), 5(Cyprus)
Matching:	1(I), 2(E), 3(M), 4(O), 5(J), 6(H), 7(B), 8(L), 9(N), 10(D), 11(K), 12(G), 13(A), 14(F), 15(C)

Chapter 2A&B

Multiple-Choice:	1(a), 2(e), 3(b), 4(d), 5(c)
True-False:	1(false), 2(false), 3(false), 4(true), 5(true), 6(true)
Fill-Ins:	1(Volga), 2(Alaska), 3(centrifugal), 4(St. Petersburg), 5(Kuzbas)
Matching:	1(D), 2(F), 3(A), 4(G), 5(H), 6(J), 7(B), 8(C), 9 (E), 10(I)

Chapter 3A&B

Multiple-Choice:	1(d), 2(a), 3(b), 4(b), 5(a), 6(c)
True-False:	1(true), 2(true), 3(false), 4(true), 5(false), 6(false)
Fill-Ins:	1(Boston), 2(Cascade), 3(St. Lawrence), 4(Northern Frontier), 5(Rocky Mountains), 6(Nunavut)
Matching:	1(G), 2(I), 3(K), 4(M), 5(O), 6(B), 7(D), 8(C), 9(H), 10(A), 11(J), 12(L), 13(E), 14(F), 15(N)

Chapter 4A&B

Multiple-Choice:	1(d), 2(a), 3(c), 4(a), 5(c), 6(e)
True-False:	1(true), 2(false), 3(false), 4(true), 5(true), 6(false)
Fill-Ins:	1(Spain), 2(*caliente*), 3(Mexico), 4(Cuba), 5(Honduras), 6(Panama)
Matching:	1(O), 2(K), 3(L), 4(N), 5(C), 6(F), 7(B), 8(M), 9(E), 10(J), 11(D), 12(H), 13(G), 14(A), 15(I)

Chapter 5A&B

Multiple-Choice:	1(b), 2(a), 3(c), 4(c), 5(d), 6(a)
True-False:	1(false), 2(false), 3(false), 4(false), 5(false), 6(false)
Fill-Ins:	1(copper), 2(CBD), 3(*templada*), 4(São Paulo), 5(Patagonia), 6(Peru)
Matching:	1(E), 2(D), 3(K), 4(G), 5(M), 6(L), 7(J), 8(H), 9(B), 10(C), 11(A), 12(I), 13(F)

Chapter 6A&B

Multiple-Choice:	1(e), 2(a), 3(a), 4(e), 5(a), 6(d)
True-False:	1(false), 2(true), 3(true), 4(true), 5(false), 6(false)
Fill-Ins:	1(Nigeria), 2(separate development), 3(malaria), 4(Cape Town), 5(The Congo), 6(Madagascar)
Matching:	1(F), 2(E), 3(G), 4(H), 5(I), 6(A), 7(D), 8(C), 9(J), 10(B)

Chapter 7A&B

Multiple-Choice:	1(c), 2(a), 3(d), 4(b), 5(c), 6(e)
True-False:	1(false), 2(false), 3(true), 4(false), 5(true), 6(false)
Fill-Ins:	1(Cyprus), 2(Relocation), 3(Shi'ah or Shi'ite), 4(Jordan), 5(Cairo), 6(Caucasus)
Matching:	1(K), 2(L), 3(E), 4(D), 5(O), 6(M), 7(F), 8(N), 9(C), 10(I), 11(A), 12(H), 13(G), 14(J), 15(B)

Chapter 8A&B

Multiple-Choice:	1(d), 2(c), 3(a), 4(c), 5(e)
True-False:	1(false) 2(true), 3(false), 4(false), 5(true), 6(false)
Fill-Ins:	1(caste), 2(Coromandel), 3(physiologic), 4(Sinhalese), 5(Arabian), 6(Nehru)
Matching:	1(F), 2(J), 3(L), 4(H), 5(I), 6(G), 7(A), 8(K), 9(E), 10(C), 11(D), 12(B)

Chapter 9A&B

Multiple-Choice:	1(b), 2(a), 3(b), 4(b), 5(a)
True-False:	1(false), 2(false), 3(false), 4(true), 5(false), 6(true)
Fill-Ins:	1(Mao Zedong), 2(Kanto), 3(Hong Kong [Xianggang]), 4(Hong Kong [Xianggang] or Macau), 5(Shenzhen, China), 6(extraterritoriality)
Matching:	1(L), 2(B), 3(G), 4(M), 5(I), 6(J), 7(A), 8(N), 9(C), 10(K), 11(D), 12(H), 13(O), 14(E), 15(F)

Chapter 10A&B

Multiple-Choice:	1(a), 2(c), 3(d), 4(a), 5(b)
True-False:	1(false), 2(true), 3(true), 4(true), 5(false)
Fill-Ins:	1(Chinese), 2(protruded), 3(Luzon), 4(Hanoi), 5(New Guinea)
Matching:	1(G), 2(F), 3(H), 4(B), 5(A), 6(C), 7(I), 8(E), 9(D), 10(J)

Chapter 11

Multiple-Choice:	1(b), 2(a), 3(d), 4(c)
True-False:	1(true), 2(true), 3(false), 4(true)
Fill-Ins:	1(Sydney), 2(South), 3(Wellington), 4(Melbourne)
Matching:	1(E), 2(B), 3(A), 4(D), 5(C), 6(F)

Chapter 12

Multiple-Choice:	1(c), 2(a)
True-False:	1(false), 2(true)
Fill-Ins:	1(Melanesia), 2(Polynesia)
Matching:	1(E), 2(C), 3(A), 4(D), 5(B)

NOTES

NOTES

NOTES

NOTES

NOTES

NOTES

NOTES

NOTES

NOTES

NOTES

NOTES

NOTES

NOTES

NOTES